SUPPRESSED SCIENCE

**Radiation, Global Warming,
Alternative Health & Healing,
Censored Scientists, Alzheimers,
Global Pandemics, Grizzly Bears,
Crop Circles . . . & Much More**

BY JACK PHILLIPS

Suppressed Science:

Radiation, Global Warming, Alternative Health & Healing, Censored Scientists, Alzheimers, Global Pandemics, Crop Circles and Much More

ISBN # 0-9785733-0-7

Published May 2006

ABOUT THE AUTHOR

John J. (Jack) Phillips, SB '38, SM '40, MBA '51, Major USAR, Retired, was born in New York City in 1917. He studied Chemical Engineering at MIT and Business Administration at Harvard.

Called to active duty in January 1941 he served almost five years as an Ordnance Officer returning home, after 3 and 1/2 years overseas, on VJ Day.

He has had industrial experience in the petroleum, textile and electronics industries. He developed one of the first chromatographic instruments for analysis of petroleum products in his own company.

He was involved with the Jupiter, Redstone and Pershing Missile industrial programs in Office, Chief of Ordnance in the Pentagon and with the Nike and Anti-Tank Missile programs at the Army Rocket and Guided Missile Agency in Huntsville, Alabama.

At NASA Headquarters he was associated with the Mercury, Gemini and Apollo programs and the Office of Advanced Research and Technology from which he retired.

He is an emeritus member of both the American Chemical Society and Sigma Xi.

TABLE OF CONTENTS

FOREWORD

In a world where new knowledge is being generated at rapid rates and a multiplicity of sources bombard us with information, it is hard to discriminate between fact and fiction. Sophisticated propaganda machines operated by governments, businesses and advocacy groups, which support activities and policies for their own purposes, make discovery of the truth difficult. Furthermore, honesty and respect for the truth has diminished during this republic's two-century lifetime. The Father of Our Country, George Washington, according to reports, couldn't tell a lie. Some of our recent presidents have not aspired to that level of probity. Many school and college students admit to cheating on examinations without apparent guilt.

An informed public, according to the Founding Fathers, can be depended upon to make good decisions in the best interests of the country. A misinformed public cannot.

Misinformation has already caused extensive damage to this country. We lag behind the rest of the world in the use of nuclear power because people have not been told the facts about radiation. Our enemies are attempting to deindustrialize the country and destroy our middle class by spurious computer forecasts of future weather catastrophes. Our medical care establishment has prevented the introduction of effective low cost advances in medical science while killing hundreds of thousands of us every year with drugs.

Hopefully this book will provide readers with reliable information on currently important issues and energize them to use their collective power to change what they deem important to change. This was the richest country in the world only a few years ago. Now we are in decline. To continue this decline is not in the best interests of the People!

JACK PHILLIPS
April 2006

Above, author, journalist and researcher Jack Phillips embarked on an expedition to Greenland in 2005 to explore claims made about the global warming phenomenon. In this photograph, Phillips is seen standing in front of a six-mile-wide glacier in a region that environmental advocates claim has been destroyed by climate change.

INTRODUCTION

My first attempt to write a newspaper article came as the result of a dinner meeting with Christopher Petherick, editor of the *American Free Press*. I interested him in a scientific paper I had obtained at a meeting of Doctors for Disaster Preparedness which debunked widespread public fear of any level of ionizing radiation. Christopher borrowed my copy of the paper to read at home and subsequently suggested I write something about it which he could use in his newspaper. My submission was printed on the "letters to the editor" page of his paper but with a larger than normal headline and a yellow background. This encouraged me to additional literary efforts which resulted in over 30 articles which are included in this book. All appeared in the *American Free Press*, except for the von Pohl article, which was published in The American Dowser.

Hopefully my pro bono efforts have helped to support one of the few independent newspapers in the United States. Also hopefully, my exposure of readers of the newspaper to facts, many of which have been suppressed, will assist them in making informed decisions, particularly in the healthcare field.

Daniel Haley, former member of the New York State Assembly, told me that he believes people in this country need to become "pitchfork mad" in order to overcome the inertia and intransigence of the medical establishment. I think that he is right and I am attempting to do my share, as he did with his book "Politics in Healing," to bring to public attention the facts concerning the consistent suppression of advancements in medical science by that establishment.

Medical care costs much too much and results in too many deaths, injuries and diseases. Technologies required to reduce costs and casualties already exist, but there are formidable barriers preventing their utilization. Unwise laws and regulations stand in the way of progress. People need less "protection" and more freedom of choice!

In 1921, Madame Marie Curie, Nobel Laureate, came to the United States as part of a a six-week tour that included a reception at the White House. There, President Warren Harding presented her with a gift of a gram of radium that cost $100,000. The photograph above is from the archives of the Historical Society of Western Pennsylvania, taken during her visit to the Standard Chemical Company's Canonsburg plant.

CHAPTER 1

IONIZING RADIATION AND ITS EFFECTS

The discovery of radioactive materials and experience with their effects generated respect for their dangerousness. Madame Curie who discovered radium died of cancer. The ladies who painted dials with radium paint in the automotive industry became casualties. Dentists lost their fingers as a result of holding photographic plates against their patients' teeth. Scientists were injured and at least one died of radiation damage during development of our atomic bomb. And the devastation of Hiroshima and Nagasaki by atom bombs generated an unreasoning fear of radiation, bolstered by politically modified science, which still interferes with its beneficial use.

The "Ghost Town" website on the internet provided the misinformation that the area around Chernobyl won't be safe for 600 years and that 300 to 300,000 people died from radiation sickness in the reactor accident there. In fact, the ghost town of Pripyat's gamma radiation dose was found to be 90 mrem per year by Polish scientists in 2001 - the same level as in Warsaw and lower than that in Grand Central Station in New York City. Furthermore the reactor disaster resulted in only 28 deaths from radiation and 3 from other causes and acute radiation sickness in 134 of the 237 reactor staff and rescue workers. This is hardly a high casualty rate for catastrophic destruction of a poorly designed reactor with 10 days of emissions of radionucleides into the atmosphere.

Not so well known is the effect of long-term exposure of apartment dwellers exposed to more than 1500 mrem per year due to accidental use of radio cobalt (Co-60) in the steel used to construct their building. They experienced a death rate of 3% of that expected for the general public and only 6.5% of expected congenital malformations in their children.

Our ancestors survived higher levels of background radiation than exist today when the earth was more radioactive. Our bodies require a certain amount of ionizing radiation in order to function efficiently! Many of us don't get enough of it. Much expensive mischief has been generated by the assumption that zero radiation has no effect on cancer rates including radon laws, which have been costly to homeowners.

On the other hand, the Earth's background radiation is not uniform. There are numerous locations where this radiation, probably coming from a natural fission reaction in the Earth's core, concentrates enough to cause cancer and other diseases. They can be located and avoided. Many German scientists have contributed to knowledge in this field.

Johan Walther coined the term geopathic and urged physicians to have homes of patients with degenerative diseases dowsed in his book, *The Mystery of the Divining Rod.* Dr. P.E. Dobler's book, *Photographischer Nachweis der Erdstralung*, covered his investigation of the effect of this radiation on photographic plates. Dr. V. Fritsch's book, *The Problem of Geopathic Phenomena from the View Point of Geophysics*, called for multidisciplinary research in this area. Dr. Joseph Wust and others used Geiger counters to prove the existence of gamma radiation under cancer beds in the town of Pleutersbach. Reinhard Schneider, a physicist, developed a system for dowsing research based on physical principles. His ideas are included in "Leitfaden fur Angewandte der Ruten und Pendelkunst" I in 1974 and II in 1984.

Radiation is akin to fire. It is very useful when under control, but very dangerous when out of control. Irrational fear of radiation is an expensive delusion. It has prevented us from benefiting, as other countries have, from the extensive use of nuclear power.

HORMESIS IS THE WORD

Radiation can be good for you, provided you don't get too much of it. You can suffer from radiation deprivation! Too little can leave you with a weak immune system, making you more susceptible to disease and that can shorten your life.

The word to know is "hormesis." Adequate amounts of ionizing radiation stimulate DNA damage control systems that decrease metabolic mutations of your stem cells permitting you to live longer. Radiation also stimulates your immune system giving you greater resistance to disease.

It is true that a high dose rate of radiation suppresses these biosystems with increased disease and shortened life span. But the fact that adequate ionizing radiation has the opposite effect is well documented. For example the Russians found that Eastern Urals villagers exposed to radiation from the Manyak explosion in 1957 had tumor related death rates which were substantially lower than those who had not been exposed. They also found incidence of lung cancer in Manyak nuclear workers was significantly reduced.

A Canadian study of breast cancer patients showed substantially reduced mortality rates for low-dosage exposure. A U.S. study of nuclear shipyard workers showed that those who had moderate lifetime doses had better mortality rates from all causes as well as all malignancies than non-nuclear workers.

To cap the climax, the Japanese, exposed to moderate doses of radiation in Nagasaki, are living longer than those who were there but were not exposed.

To many people, the above discussion represents heresy. However, the information is factual.

Dr. Myron Pollycove of the U.S. Nuclear Regulatory Commission says that low-dose whole body irradiation for cancer immunotherapy has been shown to be effective in rodents and in humans. Clinical trials for patients with breast, colon, prostate, and ovarian cancers and lymphomas are needed in the United States.

Successful implementation of these trials would provide a long sought for major advance in cancer therapy. Public recognition of radiation hormesis would terminate radiation phobia. Ending the enormous expenditure of many billions of dollars for needless protection from low dose radiation would also furnish funds needed for health care and medical research that includes low dose radiation immunotherapy for cancer and infectious diseases.

NEW HOPE FOR CANCER PATIENTS

Recent clinical experience in Japan demonstrated conclusively that whole body exposure to ionizing radiation can substantially increase survival rates of patients with non-Hodgkins Lymphoma.

A group of such patients, all having received standard chemotherapy treatments, was divided into two groups. At the end of 8 years, 84 percent of the group which additionally received 150 rads of whole body x-ray was still alive. Only 50 percent of those who received only chemotherapy were still alive at the end of 8 years. Furthermore 84 percent of the x-rayed group was still alive at the end of 13 years.

Cancer patients in France and Japan are now able to supplement conventional therapies with whole body radiation.

Too bad our own medical establishment has seen fit to ignore this breakthrough.

THE JAPANESE ATOMIC BOMB PROGRAMS

Two new historical accounts suggest that the Japanese during World War II had successfully developed atomic weapons and were working with the Germans to acquire the materials needed to use them against the United States.

According to Edward Behr's *Hirohito: Behind the Myth*, Yoshio Nishina was the father of the Japanese atomic bomb. Robert Wilcox's *Japan's Secret War* supports this opinion and suggests that they were close to a tactical capability.

According to *Behind the Myth*, Nishina, a distinguished Japanese physicist, started experimenting with nuclear fission in

1939 and published four papers in *Physical Review* and *Nature* on fission products of thorium and uranium. He designed and built Japan's first Cyclotron in 1940. He also discovered uranium 237.

In 1941, General Yasudo chose Nishina to head Japan's uranium bomb project. He assembled a team including Shinichiro Tomonaga, who won a Nobel Prize after the war.

The Japanese Navy also supported a uranium bomb project headed by Bunsaku Arakatsu who had studied under Albert Einstein. He built his own accelerator, determined that about 2.6 neutrons were produced from each fission of U-235 and published this in the Oct. 6, 1939, *Physical Review*.

Arakatsu lectured at Kyoto University on nuclear fission and the possibility of an A-bomb. He assembled a team which included Hideki Yakawa who became the first Japanese to earn a Nobel Prize in physics.

The Japanese Institute for Physical and Chemical Research supported these efforts as did a huge industrial complex in Hugnam, North Korea, owned by Jun Naguchi who had dammed the Fusin Chosin and Yalu rivers for hydroelectric installations which supplied this complex with a million kilowatts of power.

Since Japanese access to uranium was limited, they requested help from Germany. Germany obliged by dispatching the U-234, a 2,200-ton submarine commanded by one of their most experienced submariners, with 1,120 pounds of uranium oxide, mercury and other critically needed items.

Unfortunately for Japan, the U-234 was in the mid-Atlantic when Germany surrendered and Admiral Doenitz ordered its surrender to allied forces on May 10, 1945. When the sub was stopped, two Japanese Naval officers aboard committed suicide with sleeping pills and were buried at sea the next day.

The U-234 was boarded by a U.S. Navy crew on May 14 and Navy records show that its cargo included 560 kilos of uranium oxide for the Japanese Navy. This was said to be enough for two atomic bombs.

Japan's Secret War reports a Japanese officer's eyewitness account of an experimental atomic bomb explosion off the coast of North Korea near Hugnam on Aug. 10, 1945. On Aug. 15, "the voice of the Crane" that of the Japanese emperor, was heard by the Japanese people for the first time with the message that the fighting must cease.

Russia's capture of Japan's North Korean nuclear facilities may have shortened the time required for them to develop their own atomic bomb.

Ever since the first atomic bomb was dropped on Hiroshima, some Americans, including many of the scientists, have publicly bewailed the fact that they were involved in the making of this terrible weapon of mass destruction. But consider that, for the grace of God, the Japanese might have been the first to use an atomic bomb against the United States.

They had the well-trained scientists and engineers needed to produce a bomb. They recognized that such a device was possible. Their military recognized the potential and provided funds and priorities to speed development.

It was Japan's bad luck to lose the cargo of the German submarine which would have given them enough raw materials for their bombs. Additionally, U.S. success in conventional bombing of the Institute for Physical and Chemical Research, which destroyed both equipment and raw materials and necessitated the removal of what was left to Korea, delayed their programs.

RADIATION PHOBIA IMPEDES ADOPTION OF THERAPIES

At a meeting of Doctors for Disaster Preparedness in Colorado Springs on July 28 Dr. Jerry Cuttler disclosed that, contrary to popular belief, there are substantial health benefits to be gained by the judicious use of low-level radiation—and very little risk.

In the 1970s Harvard Medical School physicians found that low dose, whole body radiation was effective in treating non-

Hodgkin's lymphoma. More recently the French have had similar success.

As a result, a proposal for clinical trial of the therapy has been approved in Europe.

The Japanese have been studying the use of low- dose radiation for 20 years, using it for the suppression of diabetes, cancer and hypertension, as well as the moderation of psychological stress.

Irrational fear of any kind of radiation, fostered by antinuclear groups in the United States, coupled with a rational fear of the horde of lawyers who are ready to sue physicians unless they "do what everybody else does" appears to be preventing Americans from enjoying the benefits of this new medical technology.

The "Linear No Threshold" model of radiation carcinogen provides the basis for this irrational fear.

It originated when Americans were agonizing about the 150,000 to 200,000 Japanese who were killed in August 1945.

Many scientists who had helped develop these weapons felt responsible for the devastation that had been visited upon Hiroshima and Nagasaki. They felt they could not "put the genie back in the bottle," so they focused on limiting A-bomb development, testing and production.

Things might have been different if people realized that the Japanese themselves had begun A-bomb development in 1941 and were on the verge of having a tactical weapon by the end of the war, which they intended to use against the United States.

Dr. Arthur Robinson has stated that the controversy over how to project the data obtained at the very high levels of exposure in Hiroshima and Nagasaki to lower levels was settled, finally, by Linus Pauling at a meeting at the National Academy of Science.

Pauling proposed the assumption that zero radiation should have zero effect on cancer rates while increasing exposure to radiation would have a proportionate effect.

This assumption is the foundation upon which the "Linear No Threshold" theory is based. It persists in the face of a mass of data refuting it.

The fact is that our bodies need a certain amount of ionizing radiation in order to function properly, our immune systems and DNA repair systems are activated by radiation and our bodies were designed to function in a world where radiation is ubiquitous.

Background radiation levels are relatively low in the United States, ranging from about 0.1 Rem in the Gulf States and San Francisco to as high as 1.2 Rem in the Rocky Mountain States.

Much higher levels exist elsewhere without causing perceptible damage.

Dr. Z. Jaworoski, at a conference on radiation in Teheran, Iran, presented data based on UN Scientific Committee on the Effects of Atomic Radiation (UNSCEAR) figures, which disclosed average exposure levels in Sweden to be 1.8 Rem, Southern France 8.8, Kerala Beach, India, 3.5 Rem and Guarapari beach, Brazil 79 Rem.

The East Coast of India contains immense amounts of radioactive Thorium. People go to the beaches there for exposure to both the sun and radiation.

Misinformation about Chernobyl has been used to exacerbate this irrational fear.

The UNSCEAR report, "Sources and Effects of Ionizing Radiation," released in June 2000, includes an exhaustive study of the Chernobyl accident.

One hundred forty six people from 21 countries were involved.

UNSCEAR found that only 28 of the 134 Chernobyl employees who developed acute radiation sickness died from it. Two more died from fire and falling objects.

No increase above the normal incidence of cancers or leukemias was observed among the 381,000 clean-up workers who removed radioactive debris to allow the staff to continue operating the other three reactors.

Although some thyroid cancers were found, the highest incidence rate in Russia was 26.6 per 100,000 children and there were fewer in high dose areas.

In the United States the incidence rate is 13,000 per 100,000 people. It is 24,000 per 100,000 in Hawaii.

The UNSCEAR report concluded that, except for the people who were immediately involved, the radiation exposure from the Chernobyl accident was minimal and did not appear to elevate the risk of leukemia even for the clean-up workers. Nor was there evidence of other non-malignant disorders related to radiation.

Incidentally the background level of radiation of the evacuated land near Chernobyl is about 0.6 Rem—less than that experienced in the United States' Rocky Mountain region.

Also of interest is a 1985 report by the U.S. National Council on Radiation Protection, which states: "Available human data on low dose I-131 exposures have not shown I-131 to be carcinogenic in the human thyroid."

Additionally, the 1997 report of the National Cancer Institute's 14-year study of thyroid cancer stated that there was no evidence to associate thyroid cancer with the hundred A-bomb tests in Nevada.

Consequently the 1,800 "excess" thyroid cancers discovered in the Chernobyl screening cannot be blamed on radiation.

Isn't it about time for us to wake up and demand some changes in the way lawyers and physicians operate in this country to impede the introduction of important life-saving therapies?

RADIATION THERAPY

Recently, a retired Navy captain, age 81, who has Waldenstrom's Macroglobulinemia—a rare blood cancer first diagnosed in 1992—became disenchanted with chemotherapy treatments.

Learning about total-body irradiation therapy through a colleague in the Rickover nuclear submarine program, he sought this treatment in the United States.

More than a dozen medical and radiation oncologists refused him for a variety of reasons.

Finally, through the offices of Dr. Pollycove of the Nuclear Regulatory Agency he was referred to Dr. James S. Welch at Johns Hopkins Medical Center in Baltimore.

After reviewing the Harvard and Japanese data on the procedure, Dr. Welch treated the captain with centigrays of X-ray twice a week for five weeks—a total of 150 centigrays.

At the end of these treatments his spleen, which had been enlarged as result of the disease, returned to normal. The procedure, in contrast to chemotherapy, produced no discomfort.

Some months after discharge, Dr. Welch administered a booster series of the same low-dose radiation, limiting it to the spleen.

The captain was very pleased with this treatment at Johns Hopkins and stated that he hoped this procedure would allow him and others to avoid the sickening side effects and complications of chemotherapy.

Dr. Sakamoto, in Japan, has used total body low-dose radiation in conjunction with local high dose radiation treatments on an ovarian cancer patient after surgery. Fifteen doses of 10 centigrays during five weeks totally eliminated metastases.

When Dr. Sakamoto acquired colon cancer, he gave himself similar treatment after surgery to cure himself.

Back at the July 28 meeting of Doctors for Disaster Preparedness in Colorado, Dr. Edwin Zebrowski, first editor of *Advanced Nuclear Reactors*, disclosed that the generation of electrical power in nuclear power plants is widespread.

Many, many countries are using this technology.

Worldwide capacity is about 400,000 megawatts. The French are selling electrical power all over Europe. Well-run plants have a service factor of over 90 percent with less than 2 percent unplanned downtime.

As a result of the operation of these plants there is a worldwide stock of about 2,400 tons of plutonium plus about 100 tons of weapons-grade plutonium in the U.S. and an equal amount in the Russian sphere of activity. Since about 8 Kilograms is sufficient for one bomb this amount of plutonium creates serious problems.

Breeder reactors which can utilize plutonium as well as U238 and thorium 230 can potentially provide more total energy than all of the fossil fuels combined. They also could substantially reduce the need for radioactive material storage.

No nuclear power plants have been constructed in the United States in 30 years because of strong anti-nuclear prejudices fostered by poorly informed people. This puts us at a distinct disadvantage with respect to the rest of the world. Even Iran is building some with the help of the Russians.

It appears that the time is ripe for serious consideration of the breeder reactor as a solution to both power shortage and nuclear fuel storage problems.

GERMAN STUDIES OF NOXIOUS ENERGIES

There is a growing body of interest in the art of dowsing— using rods to find sources of water and other things buried in the earth. According to modern practitioners, the application has expanded from simply being used to locate water to finding pockets of energy, possibly radiation, emanating from so-called "geopathic" spots in the ground.

It is not widely known in the United States that cancer can be caused by emanations that come up through the ground and can penetrate several stories of buildings.

This knowledge is believed to have originated as the result of a study in the Bavarian town of Vilsbiburg in 1929. At that time Vilsbiburg had the highest cancer death rate of any town in Germany. The German government wanted to know why.

A German master dowser named Gustav Freiherr von Pohl conducted the investigation. The Burgomeister provided a police escort to ensure that the results were official and properly obtained.

Von Pohl didn't talk to people or enter houses. He did his dowsing outside. He marked, on a map of the city, those houses where his dowsing indicated residents were highly likely to get cancer.

When he was finished he gave the map to the town physician, who marked the houses where deadly cancer cases had occurred. The physician's account was in complete agreement with von Pohl's findings.

Similar tests were conducted in other towns as Mattsee, Austria and Weilburg, Germany, with similar results.

Inside the houses where cancer deaths were frequent, investigators found beds in which one person after another had died of cancer. Dowsing showed that they were located over geopathic zones where water veins crossed underneath the earth or were involved with faults in the underlying rock.

These zones appeared to be generating emanations not susceptible to detection by then-available scientific instruments.

The investigators concluded that people exposed to such emanations for long periods of time were likely to develop cancer. Spending eight hours every night over a geopathic zone was dangerous to health. However, simply moving a bed away from the zone reduced the risk. Similarly, sitting in a chair over one of these zones, at work or at home, for long periods could also be harmful.

Interestingly cats are reported to be attracted to areas where energy is noxious for humans, while dogs are reported to avoid them.

Dowsers have found noxious energies arising from TV sets, electrical power lines and computers that are electromagnetic in origin. However the nature of the emanations discovered by von Pohl is still controversial. Some believe that they are subtle energies as yet unknown to science. Dr. Josef Oberbach, author of Fire of Life, Your Bioplasma, thinks they are radioactive particles. Others are convinced that electromagnetic radiation is to blame.

In the 1930s P.E. Dobler, a German physicist, exposed photographic plates to so-called geopathic radiation and collected data that he believed could be explained as the result of electromagnetic radiation.

Jacob Stangle, a German engineer, spent 15 years developing a "dowser on wheels" using a very sensitive scintillation counter with a strip chart and electrical meters. With it he located hundreds of water wells, according to reports. A scintillation counter is a device capable of detecting and measuring radiation by counting the tiny flashes of light generated when gamma rays or charged particles impinge on a crystal sensor.

When medical researchers became aware of his device, he was asked to check von Pohl's work at Vilsbiburg. In 1972 he found sharp increases in radiation, characteristic of water veins, in three locations where von Pohl had recorded their presence in 1929.

Stangle's finding of strong radiation in areas previously known to be cancer producing is considered to be convincing evidence that pathogenic stimulation zones are real and not imaginary.

Reinhard Schneider, a German physicist, who specialized in high-frequency radio waves, also investigated dowsing phenomena. He believed that cancer is likely to be encountered above interactions of at least two water veins and/or fracture zones where the electromagnetic frequency associated with the water is in the range of 2,450 Mhz of microwave energy.

Schneider also developed special equipment and a multifaceted system for scanning the human body to identify actual and potential diseases. German law limits their use to physicians and certified healing practitioners.

Helmut Thiele of Munich, Germany, assembled a system for measuring radio frequencies over geopathic zones. He found that they are carriers for VHF and UHF radio signals and that there are anomalies at the edges of these zones, which make them dangerous for people who are exposed for appreciable lengths of time.

In studying the effects of a water vein associated with a geological fracture underneath his own living room, he found that, with his dipole antenna directed against the flow of the stream, crossing the stream from either side resulted in a voltage spike of almost 40 volts. He observed this effect at several frequencies, including some in the high Gigahertz range. He also found anomalies at different distances above the ground.

For those who would like to replicate the experiment, Thiele used a dipole antenna connected to a field strength meter and a multimeter. A laptop computer recorded his results.

It should be noted that this noxious energy, dangerous to humans, is not connected to concentrations of radon gas. There has been much concern about this substance; and, it is true, high concentrations of it can be damaging. However, in cases where normal background radiation is low, its presence in a home can actually reduce cancer risk because of the hormesis effect. Hormesis refers to the assumption that zero radiation results in nearly zero cancer risk which has been proven to be false by competent investigators.

Additional information on dowsing and noxious energy can be obtained from the American Society of Dowsers in Danville, Vt., on the web at www.dowsers.org. Dr. Ronald Blackburn is their science advisor.

FRENCH & SWISS STUDIES OF NOXIOUS ENERGIES

Many years ago, French investigators found a connection between radiation originating deep underground and cancer. Called noxious energy, it is generated in "geopathic zones" associated with water veins, faults in the underlying rock and geomagnetic grid lines.

Before World War II, Pierre Cody, a French engineer, used an electrometer to detect ions in the cellars of houses underneath more than 7,000 cancer beds in a seven year study.

In one case, he placed an electrometer directly below the cancerous growth of a woman, who had recently died, when she was lying in bed. It picked up 10 times more ions than a second device six feet away. The same result was obtained when the second electrometer was placed seventeen inches away. Cody concluded that the radiation, whatever it was, rose vertically out of the ground and did not diffuse laterally.

He placed a lead sheet under an electrometer standing over one of these noxious energy zones and found that it took 49 minutes to discharge whereas only seven seconds were needed without the lead sheet. He also found that peacock blue and canary yellow patches appeared on the underside of these lead sheets when they were left in place over such noxious zones for more than a month.

Applying this knowledge to 491 cases of illness associated with geopathic zones, he found that a lead sheet under the bed resulted in increased pain for several weeks, followed by marked improvement. Many people were healed if the lead sheets were replaced often enough. However, illnesses took a turn for the worse suddenly, after several months, if the sheets were not replaced.

Louis le Prince-Ringuet, an eminent physicist, director of the Duke Louis de Broglie Laboratory in Paris, supported Cody's study. His book, *Experimental Study of Air Ionization by Certain Radioactivity in the Soil and its Influence on Human Health*, detailed the results of this investigation. Unfortunately, the French Medical Establishment ignored it.

A Swiss nuclear physicist, Angelo Communetti, with the help of two dowsers, employees of Hoffman-LaRoche, who located water supplies worldwide for the company, confirmed some of Cody's suspicions. He found that two reaction zones, each about two yards wide, correlated with a water vein feeding a well with large quantities of water, and penetrated all floors of a five story building. The dowsing reaction had the same force on all floors even though three layers of heavily reinforced concrete separated the top floor from the bottom.

Communetti called the unidentified radiation the D-agent, where D stood for Dowsing.

In the 1960's at Moulin's in France, Dr. J. Picard spent nine years recording the location of 282 cancer deaths. Then Joseph Stangle was invited to come from Germany with his scintillation counter, Dowser on Wheels, to see what he could find.

In one house, a 12-year-old boy died from a sarcoma on the right side of his body. Subsequently a nine-year-old boy, from another family, died from a sarcoma on the right side of his body after sleeping the same bed as the as the first victim.

Stangle found radiation which he associated with water veins under the house where these boys had died. He made similar finds at many other houses where cancer deaths had occurred.

Dr. Picard never published the results of this study. He feared reprisals against his practice by the French medical establishment which was unwilling to admit the connection between noxious radiation and cancer.

BARON VON POHL'S VILSBIBURG CANCER REPORT

This article, published in *The American Dowser* in 2005, is an extract of one published in 1930 in the *Zeitschrift fur Krebsforschung*, by Gustav von Pohl, in Germany. It was translated in Germany in rough form by Helmut Thiele and I polished the English. I think it is of historical importance and may induce sensitive people with the dowsing capability to repeat Baron Gustav von Pohl's achievement.

Medical scientists do not consider the effects of noxious radiations from the earth on human beings to be important, although Gockel noted that radiation comes out of the ground and "that is surely not without influence on the human body." In 1914, Gockel was only concerned with radioactive radiations and not with the even more important negative electric radiation. Only Bach (Bad Elster) was concerned about medical effects.

The present report is the result of thirty years of observations and investigations of the influence of noxious radiations on persons, animals, trees and plants.

These radiations are found over all good underground carriers of electricity such as water veins, crystal veins, etc. They flow upward, maintaining the same width as the veins with weak side effects.

Physicists and geophysicists have not studied these radiations to any significant extent. Only a few articles have ever been published about them. Professor Blacher (Riga) is of the opinion that they come from the molten core of the earth and are modified by different materials in the earth's crust. Ambronn speaks of currents which flow from the center of the earth and show up as vertical currents in the atmosphere. Hess has identified gamma rays of radioactive substances. The existence of such radiation in the atmosphere have been found by Dr. Louis A Bauer, associated with the geomagnetism section of the Carnegie Institute, through their magnetic effects. I have found them to exist 1400 meters above the ground through measurements made in a captive balloon.

The effects of these radiations are not attenuated by ice, iron, or layers of concrete in contrast to radioactive emanations. Groundwater, brooks, and rivers provide no shield against them. Their intensity in the upper floors of tall buildings is the same as on the ground floor and the earth's surface. The boundaries of the radiation are exactly the same on the earth's surface and the top floors of buildings.

Scientists have not explored this kind of negative electric radiation. Nevertheless, it is extremely injurious when concentrated by underground carriers of electricity (water veins) and irradiating part of a person's body, whose bed or work space is directly over the source.

Sensitive dowsers know from personal experience that they suffer insomnia if, by chance, they sleep over sources of this radiation while away from home. Extremely sensitive dowsers, which I count myself to be, can actually feel these noxious earth rays if they stand or sit in them for only a few seconds. They feel a kind of electric current going through their bodies and can locate the boundaries of the radiation zones without dowsing tools.

As a result of observations made over a long period of time on the effects of noxious earth radiations on people, including myself, I found that they cause insomnia. Even sensitive persons who were not dowsers were affected. If they slept poorly at home and usually awakened tired and unmotivated but slept well away from home, invariably their beds at home stood over sources of noxious radiation. Conversely, if they slept well at home and, occasionally away from home, slept badly, without exception the insomnia was experienced in a bed irradiated with noxious energy. Other dowsers have made similar observations.

For twenty-five years, I have been studying diseases which appeared in people whose beds were over sources of noxious radiation. Without exception in all cases of insomnia, nervousness, neurasthenia, rheumatism, gout, diabetes, kidney disease, gallstones, asthma, heart disease, epilepsy, insanity, and cancer, the patient or the deceased slept in a bed over a water vein and/or the radiations connected with them.

Medical doctors with whom I have shared my investigations have always maintained that the connection between underground radiation and disease was coincidental, even though I have found no exceptions. To establish scientific proof that only certain noxious radiations caused illness, it was necessary to conduct a closed investigation of an entire town. Cancer was my first choice because medical science has not been successful in establishing the cause of the disease and also because official death certificates would provide reliable information. There was an advantage for me because all the cancer cases I had investigated occurred over very strong radiations. Consequently, I could limit my work to determining the location of very strong veins. If cancer deaths in a whole city occurred over strong veins and/or their radiations, this should prove the accuracy of all my individual investigations.

My objective was to identify all the strong underground water veins of a whole city and to mark their location on a map. This would identify houses which were over the veins. Next, this map would be compared with a map showing the houses where cancer deaths had occurred to show whether or not all cancer deaths lay exactly over the veins and/or their radiations. During the test, the "death room" had to be examined to determine whether the vein was under the room and also under the concerned bed. To the best of my knowledge, there had never been an investigation of the location of the underground water veins of a whole city before this one.

To be scientifically correct, I needed official control and a police escort to permit entering houses and to insure that my work was conducted without questioning the inhabitants. Furthermore, it had to be conducted in a city with which I was not familiar.

In December, 1928, I approached Burgomeister Brandl of Vilsbiburg in Bavaria, Germany. This city had 3300 inhabitants, 565 houses with about 900 households. For this investigation, a small city was especially suitable because the inhabitants make fewer apartment changes than those who live in large cities. Furthermore, in small cities the same families own many houses for many generations, which permits study of the hereditary aspects of cancer.

The Burgomeister understood my plan completely and agreed to a police escort and control. He also arranged for the local doctor, Obermedizinrat, Dr. Barnhuber, to produce a list of the cancer death certificates, which I would be able to view only at the end of my investigation. Unfortunately, death certificates were only available from 1918, but the officials knew of some earlier cases. There were 54 deaths; 32 males and 22 females. Names, addresses, and kinds of cancer were identified. I would like to emphasize that my work would have been impossible without the unbiased and sympathetic kindness of the named gentlemen.

My study of Vilsbiburg was completed in the seven days between the 13th and 19th of January, 1929. I was accompanied constantly by a police officer and occasionally by the Burgomeister and other gentlemen. The local general practitioner, Dr. Lifts, was also present on my first day as I entered a cancer house I had identified. There were two beds in the bedroom, one of which I had marked as standing in the radiation. The Burgomeister knew that there had been a cancer death in that house. When he questioned the inhabitants, he found that the deceased cancer patient had slept in the bed that I had marked.

Dr. Barnhuber, who doubted my theory, proposed that I drive with him to two villages where he knew of cancer deaths. In the first village, unfortunately, he showed me the house where the death had occurred. However, I was able to indicate, from outside, the location of the bedroom and the position of the bed. In the second village, I correctly determined the house with the cancer case at a distance of one hundred meters. I was also able to determine that both the kitchen and the bedroom above it were dangerous places as well as the location of the bed in which the cancer patient had slept and subsequently died.

The comparison of my map showing water veins and cancer houses in Vilsbiburg with the local physician's map of houses in which cancer patients had died was conducted by the Burgomeister in the presence of witnesses. The certified record was recorded officially. The complete record was provided to the editors of the *Zeitschrift fur Krebsforschung*, but space constraints prevented complete reproduction. However, the decisive paragraph follows:

"It is an amazing fact that all cancer deaths in Vilsbiburg were found to have occurred to people who slept above strong radiations associated with underground streams of water. If Baron von Pohl marked a house as dangerous, he was also able to predict, from outside the house, the room which was dangerous and the situation of the deathbed. In a multilevel house, he located dangerous rooms on all floors. Subsequent questioning of the descendents of the deceased by the Burgomeister, the accompanying officer, proved that the Baron's predictions were correct without exception! [ed. emphasis] He was even able to predict in which of two beds in a room the deceased had slept, to the amazement of everyone present. His predictions were accurate even in the case of a tower apartment which was twenty-two meters above the ground.

Conclusion: It is ascertained by this investigation that Baron von Pohl has proved that cancer deaths result, without exception, in houses and/or rooms and/or beds which lie above strong underground watercourses."

Although my investigation was successful, it was criticized by physicians who claimed that there were so many cancer cases in Vilsbibur, that the test was too easy. Of the 167 offices in Bavaria in the Kolbechen list, Visbiburg is the 37th worst place. Consequently, I obtained a list of cities with the lowest cancer rates in Bavaria. Grafenau was ninth in the Kolbechen list.

In Grafenau, at the request of the German Central Committee for Cancer Research, Dr. Grave, the local Bezirksartz, observed my work. There were only seventeen cancer deaths in that city since 1914. All seventeen cases had been sleeping over strong vertical noxious earth radiation. I determined this by locating underground streams associated with this radiation and without any prior knowledge of medical records of cancer deaths.

It should be noted that many individual cases have been investigated by Privy, Bach (Bad Elster), Mrs. Winzer (Dresden), Professor Dr. Wandler (attaining) in Germany and by the Capuchnin Father Randoald in Switzerland. They support my conclusions that cancer can only appear through the constant irradiation with noxious rays.

Houses and the floors of houses were considered to be factors involved with cancer by Kolb, Pinzing and Werner (Heidelberg). Kolb suggested that it might be profitable to investigate not only the dampness of floors of houses, but also the ground underneath, although he had not considered underground streams specifically. His suggestion, insofar as it includes water running underground, was correct.

Subsequently, a number of domestic and foreign scholars

related cancer to general dampness underground. High levels of groundwater were considered to be responsible for cancer; however, this was an error. I have found that groundwater neither radiates nor absorbs noxious energy. But a strong vein of flowing water underneath a deposit of groundwater under a house is able to cause cancer.

The intensity of flow of an underground stream is the major factor in determining the dangerousness of the associated noxious radiation. The depth and with of the stream are minor factors. An underground stream flowing under high pressure is always dangerous to people. A leisurely flowing stream is much less harmful.

SCIENTIFIC STUDIES OF DOWSING & WARTIME APPLICATION

Recently, *The Wall Street Journal* reported on efforts to locate tunnels being dug by the North Koreans underneath the Demilitarized Zone by employing a dowser using simple wire rods. The tone indicated that the reporter wasn't convinced it was useful.

As reported in Christopher Bird's book, *The Divining Hand*, this art was demonstrated in 1913 when Armand Vire conducted a test in connection with the Second Congress on Experimental Psychology.

Vire had special knowledge of a vast network of quarries under Paris and a unique map of them safely locked up. In testing a number of dowsers, he found that several of them were able to find underground structures that only he knew about. One of the individuals actually found a previously unknown structure.

Vire published the results in an official report to the French Academy of Sciences in *La Nature*.

He stated: "Convictions as firmly rooted as mine are not given up without excruciation. But the facts stared me in the face and I was forced to proclaim, urbi et orbi, that the dowsing ability was real and there was just cause to take dowsing seriously."

Vire's opinions were shared by some of France's most emi-

Louis J. Matacia, a former Marine who served during the Korean War, is a surveyor licensed in six states. A Trustee of the American Society of Dowsers, he has located over 1,000 water wells. In his spare time he helps find missing persons and engages in treasure hunting. During the VietnamWar, on his own initiative, he taught Marines how to dowse the location of underground tunnels and booby traps thereby saving many hundreds of lives.

nent scientists including Berthelot, d'Arsenval and Deslandres, developer of the spectroheliograph, who became president of the French Academy of Sciences commission to study dowsers.

At the same time Charles Richet, a Nobel laureate, declared: "We must accept dowsing as fact. It is useless to work up experiments merely to prove it exists. It exists. What is needed is its development."

More recently on May 12, 1968, at Camp Lejuene, N.C., six professional dowsers, led by Louis Matacia, demonstrated to Marine officers at the Counter Guerrilla Warfare Command that a simple wire hanger could be used to make a device capable of locating tunnels, hidden caches of weapons and supplies and also hidden personnel and trip wires.

A Marine colonel started out by saying that dowsing was no better than witchcraft and the demonstration would be wasted time.

Matacia met the challenge by determining the direction in which a howitzer on the marine base was pointing and how far away it was. After this he had the group's rapt attention. At the end of the day all of participants were able to get a dowsing reaction even though statistics indicate that only one-third of all people can dowse.

Knowledge of a previous Matacia demonstration for the Marines had spread to Vietnam. The Observer, published by U.S. Forces in Saigon, reported on March 13, 1967, that "Matacia's rudders" (dowsing rods) were used by Marines during the final three days of Operation Independence, three miles west of An Hoa.

Hanson Baldwin of *The New York Times* reported that the C.O. of the 13th Marine Battalion, 5th Marine Division at Camp Pendelton had demonstrated dowsing techniques to a group of officers. The officer himself said he had tried the coat hanger rods and found a tunnel whose location had been unknown to him.

He said that, despite the fact that Matacia's rudders had been demonstrated at both Quantico and Pendelton, they had yet to be officially adopted by the Marine Corps but their use was spreading.

Marine engineers at Pendelton swore by them even though they knew no more about how or why they worked than did academics or intelligence experts.

One of Matacia's attempts to generate interest resulted in a letter from the Office of the Chief of Research and Development, Department of the Army.

It had run its own investigation of Matacia's technique and found it unacceptable because "there is a low probability of making useable interpretations of the rods in an unfamiliar situation." and because "confirming criteria cannot be established to inform the operator that the method or the rods themselves are functioning properly or accurately."

Ultimately the commanding general, Marine Corps Development and Educational Command, decided, in essence, that since the scientific basis of dowsing could not be determined it was impractical to formulate doctrine, organizations and techniques in the form of military publications.

Therefore, the corps would have no interest until it could be conclusively proven that the average Marine could employ the technique with reliable results.

Most scientists have great difficulty believing that dowsers can find underground sources of water, oil, and other minerals and tunnels because there is no accepted theoretical basis for the art.

However, Dr. Zaboj V. Harvalik and Dr. Elizabeth Jurka have provided pieces of the puzzle which may, ultimately, make it acceptable to the scientific community.

Harvalik, a former member of the von Braun team, a physicist and former advisor to the U.S. Army's advanced Materials Concepts Agency, has concluded, after extensive investigation, that human beings are living magnetometers with incredible sensitivity, able to react to a magnetic gradient change of millimicrogauss. He characterizes dowsers as biophysical "magnetomer-gradiometers."

He has also found by experiment that the kidney area and the area of the pituitary gland are important to the dowsing function. If they are shielded from magnetic fields, dowsers cannot dowse.

Jurka, a psychiatrist who has studied the brainwaves of a wide variety of people, found that dowsers are unusual in that, when operating, all four levels of brainwaves were active. In simple terms their brains were functioning as if the dowsers were awake and asleep at the same time.

Harvalik believes that most people can learn to dowse but they have to learn how and then practice—a lot.

The American Society of Dowsers sponsors both beginners and advanced classes. For information write P.O. Box 24, Danville, Vt. 05828 or visit the web site at www.dowsers.org.

CHAPTER 2
GLOBAL WEATHER OUTLOOK

Current efforts to involve the United States in the requirements of the Kyoto Protocol, which calls for limitation on the production of Carbon Dioxide emissions, have been resisted so far. However, they are continuing and a serious rift has developed between the so-called "mainstream climatologists," who favor adoption of the Kyoto Protocol, and a considerable number of scientifically trained Americans. About 17,000 of them signed a petition to the President and the Congress, which recommended that the U.S. not adopt the Kyoto Protocol.

The mainstream, connected with the IPCC, forecast catastrophic effects of global warming based on the predictions of a complex computer program developed by Michael Mann et al. Those who disagree note that this program uses approximately 5,000 assumptions, is unable to account for the effects of clouds, and is inconsistent with historical weather statistics. Furthermore the validity of the program itself has recently been challenged.

If the Kyoto Protocol is adopted, it could have a devastating effect on the economic well being of our People. This country has already seen many of its industrial jobs disappear and more of them are likely to follow if Carbon Dioxide emissions are curtailed. Since water vapor is the major greenhouse gas, being responsible for 98% of the effect, some wonder whether scientific fraud might possibly be involved in the Global Warming propaganda.

GLOBAL WARMING OR GLOBAL COOLING?

There is a dispute quietly raging among scientists as to whether temperatures around the world, on average, are either warming or cooling. For the past decade, hotheads in the mainstream scientific community, backed by the well-funded environmental lobby, have been crying that the sky is falling as a result of global warming. But, in recent years, the number of cooler heads is growing. It's only a matter of time before we know who is right.

In 2004 the Russians accepted membership in the World Trade Organization in return for ratifying the Kyoto Protocol despite believing that global warming is an unverified hypothesis and probably a non-problem, according to the Access to Energy newsletter. They stand to gain billions of dollars from the sale of CO2 credits. If it takes effect here, the costs to Americans will be substantial.

The chief climatologist of the Russian Academy of Sciences, Yuri Izrael, has estimated that stabilizing the Earth's climate would cost $18 trillion. In the Feb. 18 issue of *The Wall Street Journal*, it is suggested that Kyoto could cost up to $150 billion yearly.

Rajendra Pachouri, chairman of the UN's Intergovernmental Panel on Climate Change (IPCC), recently stated that the atmosphere contains "dangerous levels of CO2," which must be reduced immediately.

On the other hand, William Ruddiman, the professor emeritus of environmental sciences at the University of Virginia, claimed in the January 24 issue of *The Irish Examiner* that, according to his climate model, the world would be one-third of the way toward glacial temperatures if it weren't for the prevalence of greenhouse gases.

Climatologist Michael Mann's computer program, which produced for the IPCC the "hockey stick chart" predicting catastrophic warming, has been accused of producing "hockey stick charts" from random data.

Many people do not realize that, in 1975, mainstream climatologists were convinced that our present interglacial period, characterized by warm weather, was coming to an abrupt end around the year 2000. They were so sure that they notified heads of state in several countries of this forecast.

Furthermore there were reports, by experts in the National Academy of Science and the CIA, which fully agreed with the mainstream and even predicted that it would be too cold to grow wheat in Canada.

Despite the overwhelming evidence, inexplicably a political decision in 1977 made funds available for research on global warming but not for global cooling.

Dr. Rodger Revelle, one of former Vice President Al Gore's professors, was involved with a National Academy of Science report which forecast global warming.

Considering the high stakes in the Kyoto political game, a review of what inspired the mainstream in 1975 to believe that the Earth was not warming, but cooling, would seem to be justified.

Willey Ley's book *The Drifting Continents* notes that three boulders about 12 miles from the center of Berlin, Germany, triggered investigations that resulted in the discovery of ice ages. The largest of these was 28 feet high and 76 feet in circumference. They were sitting on a layer of sand about 300 feet deep and there were no outcroppings of rock nearby. Similar, less impressive boulders were scattered around northern Germany on top of sand deposits. Studies in the 19th century determined that their composition matched bedrock outcroppings in Denmark.

It was at first theorized that Germany had been covered with a shallow sea and these boulders had been carried from Denmark by icebergs broken off from glaciers in that country.

However, by 1841, Swiss scientist Jean Agassiz, who subsequently became a professor of natural history at Harvard University, was able to prove that glaciers moved and carried rocks, which scored underlying structures over which they moved.

This was confirmed in 1875, when a Swedish geologist named Torell discovered that limestone deposits in Germany had been scored by glaciers.

Agassiz also found evidence of glacial scraping in rock on both sides of the Canadian border.

By 1880 most geologists accepted the idea that north Germany, Ireland, most of Great Britain and much of North America had been covered with ice. Before the end of the century, evidence of a least four glacial and interglacial periods had been uncovered.

Paleontology, the study of ancient life forms, gave birth to paleoclimatology, the study of ancient climates, around this time.

About 1945 California professor Dr. Willard Libby began using the rate of decay of carbon-14 to date organic materials. He used wood from tombs of known age to calibrate his technique. When samples from ancient European bogs associated with the end of the last glaciation were tested, they were found to be about 11,000 years old.

The glaciation process involves the removal of water from the sea and its subsequent deposit as ice in the arctic regions. The fact that there are two oxygen isotopes, O-16 and O-18, provides a means for estimating the prevalence of ice. These isotopes are non-radioactive.

According to the theory, water evaporating from the ocean should contain a high proportion of the lighter isotope and there-fore the concentration of the heavy isotope in the ocean should increase as more and more ice is formed.

The shells of certain small sea creatures contain calcium car-bonate and record the ratio of the two oxygen isotopes in the ocean water at the time they were formed. Scientists have found that shells of long-dead creatures, obtained from sea-bottom cores, show variations in this ratio, clearly indicating that the Earth has undergone many glaciations.

Dr. Fred Ward of IBM stated that a rising level of CO_2, with a characteristic waveform, had been associated with transitions to glacial conditions 17 times in the last 17 million years.

Studies of ancient peat bogs and soils confirm that the world's weather has experienced many changes over its long history. There is evidence that glacial periods of about 90,000 years have been followed by interglacial periods, never longer than 12,500 years, for the last 20 million years.

A few years ago, a study of a Chinese peat bog disclosed a

2,000-year secular down trend. The peak temperature of the last cycle occurred around A.D. 1000, corresponding to the "Golden Age of Northern Europe."

The low, about 1700 A.D., corresponded to the "Little Ice Age." Since 1700 A.D. the Earth has been warming.

However, these historical periods are not recognized by the IPCC.

Many astrophysicists believe that variations in the activity of the Sun and the orbit of the Earth around the Sun continue to be major influences on the Earth's weather as they have been for untold eons.

Another source of heat that can influence the Earth's weather is believed to be internal radioactivity. Undoubtedly this has diminished over the billions of years that the Earth has existed, but intermittent volcanoes such as Mount Saint Helens in this country are reminders of its existence.

When Karakatoa blew up in 1883, it produced a year without summer in the United States, massive local destruction and a deadly tsunami akin to the December 2004 event. The blanket of volcanic ash that it blew into the upper atmosphere cut down on the amount of sunlight that reached the Earth's surface.

Some scientists believe buildup of ice in the polar regions puts excessive pressure on the Earth's mantle, squeezing the underlying liquid magma and fostering earthquakes and volcanic eruptions. Maybe this is why Karakatoa's volcanic cone is rebuilding, as is Mount Saint Helens—perhaps for repeat performances.

Over 90 percent of the world's ice is located in Antarctica. Most of it is in a cap over the continent, which is about one and a half times as large as the United States. Some is in sea ice that surrounds it. About 5 percent is in Greenland. Again most is in an ice-cap covering most of the land and some in sea ice. Ice also floats on the ocean that covers the North Polar Region.

After 300 years of global warming, and probably at the peak temperatures for this weather cycle, it is reasonable to expect some melting. Satellite photographs indicate that the ice cover around Greenland has been reduced. Also the Swiss are reporting glaciers melting in ski areas.

However, the fact that U.S. aircraft, ditched on Greenland, were covered with 250 feet of ice after 50 years of storage is significant. Also significant is a government report that icebreakers had to deal with 80 miles of sea ice to reach McMurdo station recently—instead of the usual 10 miles.

Extensive research in Antarctica indicates that ice is accumulating there. Radar measurements disclosed that an estimated 26.8 billion tons of ice are deposited yearly on the cap according to an article in Science in 2002.

In 2004 the science magazine, *Geothermal Letters*, carried a report that sea ice has been increasing since 1979. The peninsula has warmed recently, due to changes in wind patterns bringing in more warm air from the surrounding sea. However, the interior of the continent has been cooling.

In 1999, after studying 420,000 years of climate history, renowned climatologist J.R. Petit reported that the Earth was warmer during the last four interglacial periods than it is in this one. Other reports indicate that less ice has melted in this interglacial period than in the last one.

In the light of these facts, the opinion of Dr. Richard Lindzen, a climatologist at MIT, is of interest. Lindzen said: "the catastrophic consequences of global warming, if it should occur, are almost completely speculative." Perhaps they can also be considered to be unlikely.

ASTROPHYSICIST PREDICTS GLOBAL ICEHOUSE

At an annual meeting to study disaster preparedness, a noted scientist told a group of medical researchers that, contrary to claims by the environmental lobby, the Earth may be facing another dramatic climate change in the years to come similar to the fabled Ice Age that occurred eons ago.

Dr. Willie Soon of the Harvard-Smithsonian Center for Astrophysics, at a meeting of Doctors for Disaster Preparedness in Colorado Springs, Colo., on July 27 predicted that the Earth will be an orbital icehouse 100,000 years from now.

It is unlikely that any of us will be around then, but his presentation indicated that the Earth is more likely to get colder than it is to warm up in our lifetime.

The moon, the sun and the planets exert gravitational forces on the Earth, which affect the weather.

These forces have three primary effects.

First, the Earth wobbles which affects the equinoxes about every 21,000 years.

Secondly, the tilt angle changes between 22 and 25 degrees with an approximate 41-year cycle.

Thirdly, the eccentricity of the Earth's orbit—its deviation from a circle—experiences one cycle of about 100,000 years and another of 400,000 years.

It is the combination of these effects that cause the distribution of sunlight on the Earth's surface to change from year to year.

Milutin Malankovic, A Serbian scientist, developed a theory to explain how orbital changes affect the weather.

He proposed that changes in the seasonal distribution of sunlight cause changes in Earth temperatures.

Malankovic claims that when high northern latitudes receive less than average heating from the Sun during the summer, ice from the previous winter tends to persist. Since ice and snow reflect sunlight, less heat is absorbed thereby progressively promoting additional ice formation. The more ice that is formed, the less heat is absorbed and the colder it gets.

What has been sadly lacking in the current discussion about "Global Warming" is a picture of where Earth temperatures are now in relation to where they have been over a significant period of time.

One picture, courtesy of BGR in Germany, shows that the

Earth gets hot every 130,000,000 years or so, and that warm periods do not last long. The picture reveals that we are not in a warm period now.

Another picture shows the varying temperatures at four different locations during the last 1,000 years together with atmospheric carbon dioxide concentrations.

These do not demonstrate the kind of relationship between carbon dioxide concentration and temperature that the proponents of the Kyoto Protocol say exists. Perhaps this is because carbon dioxide is only a minor greenhouse gas.

When Dr. Joel Kauffman performed infrared absorption measurements on Philadelphia air in 1999, he found that water vapor was responsible for 92 percent and carbon dioxide for only 8 percent of the absorption.

It is well known that weather prediction is a very difficult art and that much of it is based on assumptions. It is said that the latest computer programs for weather forecasting contain over 5,000 assumptions.

In view of that fact it may be sensible to wait and see if another "Golden Age of Northern Europe" develops, to see if the Earth warms to temperatures that existed in 1,000 A.D. instead of the disaster forecast by those who are politically modifying science.

WEATHER HISTORY AND GLOBAL WARMING MYTHS

New information prepared by a German weather institute could finally dispel politically motivated half-truths that the Earth is destined for catastrophe as a result of global warming.

Four charts, prepared by BGR, a German weather research institution, and published in AFP, show definitively that the Earth's temperatures have varied considerably over time, rising and falling as part of a natural cycle—influenced little by modern man or industry.

Figure 1 shows that temperatures in Greenland were much

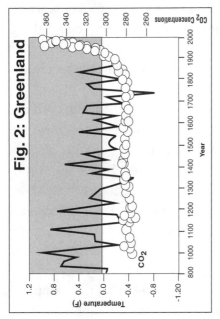

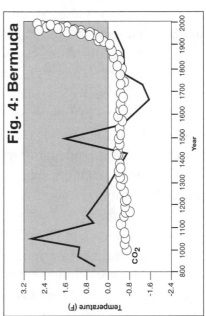

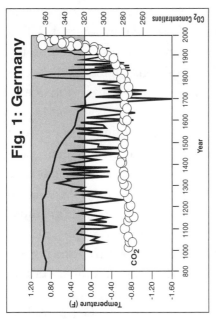

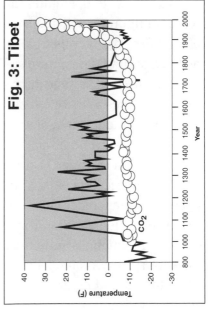

higher than usual between AD. 800 and 1000.

This was the period when Vikings from modern day Norway and Sweden engaged in long voyages of discovery, establishing colonies in Greenland and beyond.

Between 1250 and 1350, as shown on the chart, temperatures were much lower.

There is an historical record which relates that the Viking colonists were forced to leave Greenland about 1300 because the low temperatures made farming impractical.

Still lower temperatures were experienced in A.D. 1700, called the Little Ice Age. Correspondingly, the chart records show low temperatures in A.D. 1700 to 1750. Since A.D. 1700, the chart shows temperatures in Greenland trending upwards.

Evidently global warming has been happening during the last 300 years. It is not something that started recently.

In figure 2, the German experience also shows that temperatures were very low in A.D. 1700 and that they are higher now.

Very low temperatures appear about 1725 in the Tibetan record, in figure 3, and about 1670 in Bermuda, in figure 4. These data confirm the fact that global warming is not a new thing, but has been around for centuries. The lines constructed from circles on these charts represent the concentrations of carbon dioxide in the atmospheres at these locations.

Environmentalists argue that there is a connection between rising temperatures and increasing so-called greenhouse gases such as carbon dioxide. These gases, they claim, would also have a negative effect on the Earth. But quite obviously there is no apparent rise in temperature accompanying the rise in carbon dioxide concentrations.

The fluctuations in temperatures which occurred during the period of rapidly rising concentrations of carbon dioxide were all within the normal range.

In fact, quite the opposite has occurred. The German chart shows that when temperatures were higher in 1790, carbon dioxide concentrations were low. In 1990, when temperatures dropped, the concentrations were high.

The facts make it clear that the warm temperatures we are experiencing now are the result of 300 years of natural trends.

Perhaps all should pray that this warming trend continues because, as the charts indicate, there is a distinct probability that the Earth is going to get much colder in the future.

HURRICANE PREDICTIONS

Claims that global warming is behind increased hurricane activity in the Caribbean are pitting science against the politics of environmentalism. Another scientist has jumped off the government ship following a statement from the leading government panel on climate, which linked man's purported effects on the environment with heightened storm activity.

The May-June 2005 edition of *The American Scientist* reports that Christopher Landsea, an expert on hurricanes with the National Oceanic and Atmospheric Administration in Miami, FL, has resigned from the Intergovernmental Panel on Climate Change (IPCC).

Landsea's, resignation followed a press briefing by Dr. Kevin Trenberth, the lead author responsible for preparing the text on hurricanes for the 2007 IPCC report.

Trenberth, of the National Center on Atmospheric Research in Boulder, CO, told reporters that there was no guarantee that the active 2004 hurricane season would be repeated because of the natural variability in hurricane activity.

However, he suggested that a longer-term trend associated with global warming would be superimposed on that variability. Landsea's open letter to the climate science community concluded that this public statement by the IPCC's point man on hurricanes was so far outside of current scientific understanding that he questioned his objectivity.

Hans von Storch, a climatologist at the Institute for Coastal research in Geesthatch, Germany, said that Trenberth's statement indicates how highly politicized the IPCC process has become.

Storch and Lars Barring of Lund University in Sweden studied the storms in Scandinavia from Napoleon's time to the present and found no overall trend for this entire period. He and five coauthors, in a paper published in Science magazine last year, also suggested that climate variability in the past was greater than many IPCC experts estimated.

Landsea believes that hurricane seasons in the next 20 years will be much the same as those experienced in the last 10 years. The influence of global warming on hurricanes, if it exists at all, is not proven. Increases in wind speeds predicted by the computer models are not substantial.

Is the past success of efforts to blame all kinds of meteorological effects on miniscule increases in CO2 concentration, measured in parts per million, a tribute to the propaganda expertise of scientists and politicians who do not mind hoodwinking a gullible public to feather their own nests?

MIT meteorology professor, Dr. Richard Lindzen, in an article in the magazine Regulation, stated that water vapor is responsible for more than 98% of the so-called "greenhouse effect" which is supposed to have caused global warming in the last century. He noted that carbon dioxide, methane and chlorofluorohydrocarbons are minor greenhouse gases.

WATER FROM ROCKS

Water shortages are frequently in the news nowadays just as they have been off and on for over 50 years. But is it possible that water could be available in the hard rocks underneath the Earth?

Hydrologists answer this question with a resounding no. They claim that all of the Earth's water is either engaged in a continuing cycle of evaporation and rainfall, or trapped in underground reservoirs where supply is limited by the capacity of the reservoir.

On the other hand, there is evidence that their view of the world may be flawed.

On Feb. 14, 1956, the Peaty & Furman Construction and Engineering Company, excavating for construction of an addition to the Harlem Hospital, encountered a flood of water as they removed hard rock 12 feet below the first floor of the adjacent hospital building.

Investigation proved that it did not come from the Croton Reservoir, the Harlem River or the nearby sewer pipes. Hospital personnel determined that it was fit for consumption without chlorination. Its temperature increased a little at first, but leveled off at 68 degrees F independent of the weather.

Dr. Michael Salzman's article about the probable source of this 3 million gallon per day (gpd) or 2,000 gallons per minute (gpm) flood at 136th St. and 5th Ave. in New York City was published in The Engineering News Record. Salzman suggested that the water came from deep within the Earth's crust and was derived from geochemical processes occurring between rocks and minerals.

Ultimately the cement foundations of the hospital addition closed off this conveniently located supply of water when the weight of the structure was deemed sufficient to counteract the pressure of the water.

Most New Yorkers know that their city sits on a rocky island not favorably located for the collection and use of local streams. Hydrologists would not expect to find water there because it is an inhospitable place for underground aquifers. Nevertheless in 1834 there was a 100 foot deep well at 13th and Broadway which produced 21,000 gpd per day. And another at Broadway and Bleeker streets yielded 44,000 gpd from a depth of 400 feet.

The idea that water can be obtained from the Earth's rocks is not new. Aristotle in ancient Greece, Vitruvius in ancient Rome and Georgius Agricola in the Middle Ages all wrote about water coming from within the Earth. Vitruvius wrote that water was best found, not in sands, gravels and soils, but in rocks. When the Israelites needed water in the desert, Yahweh told Moses to get it out of a rock.

Salzman's *New Water for a Thirsty World* notes that there are places where rain is scarce but supplies of water are plentiful.

For example Jericho, a very ancient city mentioned in the Bible, has a veritable river pouring out of a rock for its water supply. This has been an unfailing source of water for thousands of years.

Similarly, Nefta, an oasis in the Sahara Desert, has had copious amounts of water pouring out of the ground for centuries, although rain only falls once every third or fourth year.

Where do these waters come from?

Christopher Bird's book, *The Divining Hand*, provides much information about Stephan Riess, whose theories, put into practice, benefited many people. The fascinating story of how these theories developed and how they were applied is, by itself, worth the price of the book.

Riess's interest in water from rock was initiated by a dynamite blast in a deep mine located high on a mountain. When the dust cleared he was amazed to see water coming out of the hole in quantities greater than a 25,000-gpm pump could handle. The temperature and the analysis of the water showed that it wasn't ground water. None of his textbooks was able to explain how water falling as rain in a valley could rise up almost to the top of a mountain through hard rock. Here was a mystery.

Late one night in a mine Riess said he heard a peculiar hissing sound and a trickle of water. These were the clues that led to the formulation of his theory. He observed that the mud, which usually lies under a layer of water when the mine was not operating, had formed an arch over the water and gas bubbles were rising through it. When he lit a match there was a small implosion indicating that the gas was hydrogen.

Evidently Riess concluded that minerals in the mud were reacting to form hydrogen and water.

CHAPTER 3

MERCURY AND DISEASES OF YOUNG AND OLD

A substantial body of knowledge has implicated the element mercury with several devastating diseases, which have "unknown causes." The medical establishment's approved treatments are palliative and costs of care are growing yearly.

Autism, which affects very young children who suddenly cease communicating and become excessively difficult for parents to handle, has reached a stage where it is reasonable to call it an epidemic. Autism and autism spectrum diseases, like attention deficit disorder, affect boys 4 times more often than girls. Students with these diseases substantially increase the costs of education and make the job of teachers exceedingly difficult.

In the absence of an effective research program by the governmental authorities, private citizens, particularly those with autistic children, have organized their own programs. The best known of these is Dr. Bernard Rimland's Autism Research Institute. Fortunately, they have made much progress both in identifying the cause of the disease and developing approaches to its cure.

Recent research clearly indicates that some diseases of the elderly, like senility, Alzheimer's disease, and Parkinson's disease, may also be products of mercurial toxemia. If so, Dr. Rimland's pioneering efforts may have a substantial effect on efforts to rein in our burgeoning medical expenditures.

AUTISM: THE HOMEGROWN EPIDEMIC

The New York Times of Oct. 1, 2004, records the fact that the number of children with autism or related diseases in the public school system increased from 118,846 in 2002 to 141,022 in 2003, an increase of 19 percent. It also states that this figure has increased 20 percent to 25 percent every year since the mid 1990s.

As a result of these increases, more people are required to teach these learning disabled children, which increases the cost of the public school system.

Also there has been a concomitant increase in related diseases, like attention deficit disorder, which has caused disruptions in classrooms that interfere with the education of more normal children.

The net result is that public education keeps increasing in cost and the results are far from satisfactory.

Autism was unknown in ancient times. It appeared suddenly as a uniquely American disease about 60 years ago. The first case was diagnosed in 1941. Dr. Leo Kanner of Johns Hopkins University first described it in a medical textbook in 1943. He described 11 children with "extreme autistic aloneness." All of these were born after 1930.

Originally, Dr. Bruno Bettleheim proposed that autism was caused by "refrigerator" mothers who did not provide a stimulating environment for their children. However, this theory has been discredited.

The disease, which affects boys and girls in a four to one ratio, has reached epidemic proportions in the United States. During the 10-year period ending in 2002 there was a 1,700 percent increase in the number of autistic children in American schools. The total number of disabled children in school increased also, but only by 30 percent.

The Individuals with Disabilities Act was passed in 1975 to ensure equal educational opportunities for disabled children. At that time Congress promised to cover 40 percent of the costs associated with the program, but it has never funded more than 15 percent according to Dr. Yazbak who provides a perspective on autism in the winter 2003 issue of *The Journal of American Physicians and Surgeons*.

An autistic child can cost a school system $30,000 a year or more. Since it has been progressively more difficult to fund this program, diagnosis is a serious matter for physicians and not easily approved by school authorities. Consequently there are probably more autistic children than those in the statistics. The overflow probably appears as autism related diseases like attention deficit disorder. Such children are usually disruptive in regular classrooms and, as a result, are usually controlled with drugs. Nevertheless they often interfere with the education of their classmates.

At the June 2004 meeting of Doctors for Disaster Preparedness in San Diego, California, Dr. Boyd Haley, Chairman of the Chemistry Department at the University of Kentucky, disclosed that boys are no longer superior to girls in mathematics and science. Boys have also lost about 100 points in test scores. He said affirmative action is needed to get adequate numbers into law and medical school.

Haley believes that the probable cause of this flip-flop in intellectual capabilities is mercury poisoning, which causes something like intoxication. He calls it Mad Child Disease.

In Haley's laboratory experiments, testosterone potentiated the toxicity of mercury and estrogen suppressed it. Haley said he believes this is why boys are affected so much more than girls.

Mercury is an extremely effective neurotoxin. Mercury from amalgam fillings in a mother's teeth is transferred to infants during gestation. The more fillings the greater the transfer.

Amalgam fillings were introduced into dental practice in this country around 1860. By about 1880 a uniquely American disease called neurasthenia, American nervousness, appeared. Relief of severe neurological conditions by removal of amalgam fillings was reported over 100 years ago in this country and 75 years ago in Germany. Women with more than 10 amalgam fillings and a high concentration of testosterone in their amniotic fluid are likely to have an autistic child.

Thimerosal, a preservative used in vaccines, is another source of mercury. In 1977, a report in the medical literature disclosed that a 1 percent solution of thimerosal, used as a topical disinfectant, had caused the deaths of 10 out of 13 infants. The dead infants had high concentrations of mercury in their organs. Those who survived had high concentrations of mercury in their blood.

Even though the article recommended banning this material from hospitals, this same substance, which killed children when applied to their umbilical cords, has been injected into newborn babies in this country since the mandated vaccination program started about 1985.

The amount injected is said to be 50 to 100 times as much as the Environmental Protection Agency says is a safe level for mercury taken in through the mouth.

Haley's laboratory studies show that a substance called tubulin, essential for the proper functioning of neurons in the brain, is denatured by thimerosal, which has been used in medications since about 1930.

Thimerosal disappears quickly after injection because it dissociates in the presence of water and salts in the body to form ethyl mercury. This is a far more toxic substance than mercury.

In addition, the vaccines contain aluminum, which synergistically makes it even more toxic. Furthermore lead also increases the toxicity of mercury and this is still being leached from old pipes in many inner city locations. Antibiotics also potentiate mercury.

HEART ATTACK CONNECTION

Idiopathic dilated cardiomyopathy, which has claimed the lives of a number of athletes, who dropped dead for no apparent reason, and caused a number of older people to require heart transplants, also has a mercury connection. German investigators have found high concentrations in damaged hearts—22,000 times the usual level of mercury in heart muscle. The cardiovascular systems of autistic infants are also adversely affected by mercury.

Some people are able to excrete mercury fairly easily. After exposure they have relatively high concentrations in their hair, blood and fingernails. Others, less favored by genetics, have lower concentrations in hair, blood and fingernails.

Autistic children have great difficulty eliminating mercury. They have higher than normal concentrations of mercury in their baby teeth. Mercury finds a home in their tissues from which it is not easily removed.

Since the brain constitutes about 30 percent of a newborn baby's weight, it is reasonable to assume that a substantial amount of the ethyl mercury enters it when babies are inoculated with vaccines.

Babies get a dose of hepatitis B vaccine on the day they are born. This contains 12.5 micrograms (mcg) of mercury. A 6.6-pound (3 kilograms) child takes in approximately 4.2 mcg of mercury per kilogram—42 times the allowable oral intake (0.1 mcg/kg/day) according to CDC and FDA.

It is interesting to note that thimerosal was eliminated from vaccines for cats and dogs in 1992; the same year it disappeared from vaccines in Denmark. But thimerosal-free hepatitis B vaccine for infants was not available until 2000 in the United States.

Haley believes that the United States is experiencing an unprecedented disaster as a result of a general mercurial toxemia, which is being ignored by medical authorities. He has been concerned about it and has attempted many times without success to get the FDA and the vaccine branch of the CDC to recognize this during the past 15 years.

The November 2004 University of California (Berkeley) Wellness Letter advises its readers that "all is well in Vaccineland." Thimerosol does not release mercury in the body. People are unnecessarily concerned.

Is this evidence that Haley is right when he suggests that many good doctors are dumb chemists?

Dr. Bernard Rimland, founder of the Autism Research Institute, disclosed that, when his autistic child was born in 1956, it was autistic at birth, and autism was a rare disease. It was first described in a textbook in 1943. Today, it affects 60 out of 10,000 children and most autistic behavior starts at 18 months instead of at birth. In other words, the autism occurs subsequent to the vaccinations.

A rise in the number of autistic children with little or no mental retardation has been observed. In 1987, 19 percent were in this category and 56 percent are now.

Although medical authorities claim that there is no cure for autism, Rimland stated that removal of mercury from the bodies of autistic children, by use of a special chelating agent, was effective about 70 percent of the time.

The fact that mercury is poisonous was known in ancient times. There is a legend of a king bothered by excessive hair, who for a time employed a stranger to his country who claimed to be expert in removing hair. However, when his counselors told the king about the mercury compound that the stranger was using, the king had him executed.

In England, years ago, such compounds were used to remove hair from skins in the manufacture of hats. The hatters developed mental problems memorialized by the Mad Hatter in Alice in Wonderland.

In more recent times mercury was implicated in pink disease, which affected babies between six months and two years in the English-speaking world and two to five years in Europe. The symptoms included pink fingers and toes accompanied by irritability, weakness, the sudden and rapid racing of the heart, high blood pressure, photophobia and inflammation of the peripheral nerves also known as polyneuritis.

One mother is reported to have said: "My child behaves like a mad dog!" Popular theories blamed viruses and nutritional deficiencies, but by 1950 mercury compounds, widely used in teething powders, became suspect. The disease disappeared after 1954 when mercury was removed from teething powder.

In 1928 Eli Lilly discovered thimerosal. It was patented in 1930 and introduced into various medications as a preservative. In

1941 the first case of autism was encountered. All of the 11 cases covered in the 1943 textbook, which described autism, were born after 1930. Rimland thinks that this is more than a coincidence.

The fact that Kuwait's autism epidemic started after they adopted the World Health Organization's recommended vaccination schedule is additional support for the conclusion that something in the vaccines is causing the problem. Despite the enormous amount of pollution resulting from hundreds of burning oil wells at the end of the Gulf War, there were few birth defects and no autism encountered at that time.

Haley's laboratory data clearly indicate, but do not prove, that the ethyl mercury carrier, thimerosal, the preservative in vaccines, is responsible for an epidemic of autism related diseases which has damaged a whole generation of our young.

A decision to remove thimerosal from pediatric vaccines in the United States was made in 1999. The American Academy of Pediatrics claims that all routinely recommended infant vaccines are free from preservative thimerosal. However Drs. Mark and David Geier reported that the 2003 Physicians Desk Reference reported that Merck's pediatric hepatitis B vaccine contains 12.5 mcg of thimerosal per dose and the adult version contains 25 mcg per dose. Wyeth and Aventis Pasteur vaccines also contained this preservative. All influenza vaccines contain it, and they are being recommended for children.

Package inserts are labeled per FDA regulations, and it is a criminal offense to mislabel products. Yet a spokesman for Aventis Pasteur told the Geiers that the company had ceased selling DtaP, a pediatric vaccine, in the preservative formulation in March 2001 and that their 2003 version's package insert, which listed thimerosal, was incorrect.

The World Health Organization believes that removal of thimerosal from vaccines to be unlikely because it is needed to kill bacteria during the nonsterile manufacturing process. It may also be an integral part of the process and not intended to serve primarily as a preservative. Consequently "preservative free" vaccines may not be free of thimerosal.

The Children's Health Act of 2000 established an Interagency Autism Coordinating Committee with a 10-year goal of preventing 25 percent of cases of autism through early intervention and developing better ways to treat it. Complete prevention, however, does not seem to be the objective.

ALZHEIMER'S DISEASE

According to Dr. Haley, Chairman of the Chemistry Department of the University of Kentucky, some of whose publications can be found in the *Proceedings of the National Academy of Sciences*, mercury, and only mercury, causes all the abnormal biochemical reactions that kill neurons and cause Alzheimer's disease (AD). He has found that laboratory cultures of neurons, when treated with mercury, form the tangles and deposits observed when the brains of people with AD are autopsied.

Since the FDA, the NIH and the Alzheimer's Research Foundation generally refuse to support research on toxic metals as a cause of disease, much of the research, on which Dr. Haley based his conclusion, was accomplished in other countries.

He claims that mercury fillings are a major source of our body burden of mercury. He finds that dental amalgam generates colorless mercury vapor at room temperature, which can be observed with the EPA's mercury sniffer. Brushing the amalgam or immersing it in hot coffee substantially increases emission. He can kill neurons in culture with water in which amalgam has been soaked.

The Alzheimer's Association, on the other hand, denies that there is scientific evidence showing a connection between dental fillings, which typically contain 50% mercury, and AD. An FDA panel is on record as concluding that amalgam fillings pose no danger.

A few years ago the AD and Chemistry departments of the University of Kentucky discovered that AD brains contained mercury and that the concentration in patient's fingernails decreased as the disease progressed indicating increasing difficulty in excreting it. Their grant was terminated on the issuance of their report. NIH subsequently funded a study by Dr. Saxe, a dentist at the same university to study the effect of dental amalgams on AD. His research was published in the *Journal of the American Dental Association.* His conclusions were that there was no significant association between amalgam fillings and AD and no correlation between levels of mercury in AD brains and AD.

Dr. Saxe's research and a review of the health effects of mercury published in the *Journal of the American Dental Association* underpin the endorsement of amalgam filling as safe by the FDA, the Public Health Service and the World Health Organization.

Dr. Robert J. Rowen's "Second Opinion" letter of April 2005 notes that vaccines also contain thimerosol, a mercury compound, and that adults getting the flu vaccine annually for 5 years have a substantially increased risk of getting AD. He also reports that transdermal applications of a special mercury chelating agent called DMPS has been found effective in helping autistic children, who are also affected with mercury toxicity.

Each dose of influenza vaccine contains 25 micrograms of mercury in the form of thimerosol, as a preservative. For someone weighing 110 pounds, this represents 5 times the amount that the EPA says is safe to take by mouth. But this does not enter our digestive system where 2/3 of the protection against mercury is located. Therefore it should have an enhanced toxic effect.

Mercury is also found in the air we breathe since it comes out of the smokestacks of coal burning power plants. Some is imported, like almost everything else, from China, which burns a lot of coal. Large quantities are said to get into the rivers in Brazil from gold mining activities. Since rivers usually empty into oceans, this is one source of seafood contamination.

AD has devastating effects on victims and their families. Besides the emotional strain there is a substantial financial burden.

At least $100 million of direct and indirect annual costs have been estimated. These are expected to increase because of inflation and increasing numbers of patients. There were about 4.5 million in 2000. In 2050 there may be 11 to 16 million.

Estimated cost of care for the 8 to 20 year duration of this illness is $174,000. Nursing homes charge $42 to 70,000 per year. Supplemental care, for the 70% who depend on their families, is said to be $12,000 per year.

Considering the importance of this disease and the accumulating evidence that implicates heavy metals in a number of other diseases, it is hard to see why, as Dr. Haley has stated, our medical establishment refuses to fund what appears to be a profitable area for research.

TIME FOR ACTION

Dr. Boyd Haley believes that the CDC and the FDA are strongly influenced by the pharmaceutical and vaccine industries and that they have been derelict in their duty to safeguard the health of the American People.

As a result of their delinquency, we have been systematically poisoned by mercury derived from silver amalgam fillings in our teeth and our children, especially boys, have been severely damaged by vaccines containing thimerosal.

Autism and the autism spectrum diseases, in which four boys are affected for each girl, varying degrees of madness due to mercurial toxicity, have had a devastating effect on an entire generation of young men. The cost of special education for learning disabled children is over $3 billion annually and increasing. A great number of them are autistic. Since such children are plagued with autoimmune and digestive diseases as well, their medical care is expensive.

Mercury is present in the brains of Alzheimer's disease patients and only it is capable of inactivating the enzymes that protect neurons from destruction and producing the tangles and plaque deposits characteristic of the disease. The costs of care for patients with this disease exceed $100 billion a year and are increasing.

In effect we have a health care disaster, worsening daily, caused by the unwise use of a dangerously neurotoxic material by dentists and physicians with the support and approval of their professional associations and health authorities at the highest levels of the Federal Government. The damage done to the World Trade Center by a small band of terrorists pales in comparison to the damage that has been caused by mercury in our dental fillings and in vaccines injected into our children and us.

"Mercury on the Brain," an article in the May-June 2004 *Harvard Magazine*, states that children in the Danish Faroe Islands are suffering irreparable damage from mercury poisoning.

Professor Philippe Grandjean at the Harvard School of Public Health has been studying mercury-exposed children for almost 20 years. He notes that the federal government took almost a decade to pass laws phasing lead, another toxic metal, out of the many products that used to contain it. He believes that there will be resistance to regulating mercury and favors quicker action to eliminate exposure because "you don't get a second chance to build your brain."

OUTLAW USE OF MERCURY BY DOCTORS & DENTISTS

This is not a time for scientific studies and congressional hearings. It is a time for action! Common sense calls for termination of use of amalgam fillings and the removal of thimerosal from vaccines now. The ponderous machinery of Washington has proven itself unequal to the task of protecting our health. The states need to follow the lead of Maine, which requires its dentists to warn patients of the dangers of amalgam fillings. But that limited action is not enough. Amalgam fillings, thimerosal and all other mercury containing substances should be illegal to use on or in people. Dr. Haley is currently acting as an expert witness in California to help ban thimerosal in that state.

People with autistic children and those with AD patients in their families need to be aware of the cause of these diseases and jointly call for the elimination of use of mercury by dentists and physicians. Perhaps state courts could be induced to grant injunctions against the use of these materials while state legislatures pursue the necessary lawmaking procedures.

TREATMENTS FOR MERCURIAL TOXEMIAS

The controversy over Autism and autism spectrum diseases has been raging for over 15 years. During that time the medical establishment has maintained that the cause of the disease is unknown. Some physicians and scientist's with autistic children and others familiar with the disease have insisted that it is caused by mercury, the chemical symbol for which is Hg. Their case is strengthened by the Agency for Toxic Substances and Disease Registry.

The Registry describes Hg, as a developmental toxin whose effects are dependent upon the level of exposure and the time of the dose. Offspring of exposed mothers exhibit symptoms, primarily neurological in origin, ranging from delays in motor and verbal development to brain damage. Infants with this disease may appear normal at birth but exhibit slow development or even brain damage leading to mental retardation, uncoordination and inability to move. Babies of mothers exposed to very toxic levels of Hg during pregnancy may go blind, exhibit involuntary muscle contractions and seizures and be unable to speak.

Dr. Haley, Chairman of the Department of Chemistry at the University of Kentucky claims that extremely low levels of Hg can cause neurological and other damage. Concentrations of 1 to 50 nanomolar (10^{-9} molar) kill neurons in culture and the presence of aluminum, lead, antibiotics and testosterone increase its killing power.

Adults exhibit a 78-fold variation in resistance to this toxin and infants and even greater variation – up to 10,000 fold. It is known to interfere with growth mechanisms. Exposure during windows of development is most damaging. The human brain develops rapidly during the first year of life and the blood-brain barrier is not fully developed for a considerable time after birth. Children with immune systems problems may be especially susceptible to autism

The medical establishment considers autism to be a psychiatric disease. However, Dr. Bernard Rimland's Autism Research Institute, formed in 1967 after his wife gave birth to an autistic son in 1956, has made considerable progress in developing effective non-psychiatric treatments for it. Physicians and scientists whose children have had the disease as well as their sympathetic colleagues have supplemented his efforts.

In December of last year physicians and scientists associated with the Autism Research Institute produced a consensus document (CD), which summarized the results of various pertinent investigations and provided guidelines for treatment.

They noted that animal studies indicate that infants do not excrete Hg until after they are weaned and that milk increases gastrointestinal absorption of metals. Also that bile production is necessary for excretion of Hg and this is often inadequate in infants. Since intestinal bacteria play a role in the excretory process, antibiotic exposure decreases elimination. Also stress and illness decrease glutathione levels in the body and this compound plays an essential role in getting rid of Hg.

Three chelation agents are now being used to remove Hg from autistic children. These are DMSA. DMPS and TTFD. TTFD is being used in an FDA approved experimental drug trial. DMSA is FDA approved for treating lead poisoning in children. DMPS is not FDA approved but has been available for years over-the counter in Germany and by prescription in other European countries. Presently DMPS may to be the most effective in removing Hg from children.

Dieter Klinghardt, MD, according to Dr. R. J. Rowen's "Second Opinion" letter introduced DMPS to the integrative medical community 15 years ago and is remarkably successful using it. His protocol includes removal of allergens and toxic foods, sublingual B-12 and folate to restore methylation and RNA products

Dr. Rashid Buttar is an expert in heavy metal poisoning. His experience with his autistic child is informative. At 16 months his son lost his 15-word vocabulary. Dr. Buttar's tests produced negative results 3 times in a row, but the 4th found Hg. His newly developed Transdermal treatment with DMPS removed the Hg and his son began to speak again after 8 months of treatment. At 42 months his vocabulary included 500 words and he became the youngest person to testify in Congress – about autism.

Dr. Rowen also mentions that Dr. Amy Yasko, a microbiologist, found that autistic children have a methylation defect which interferes with the ability of cells to utilize folate in functioning, repairing and dividing. About 20% of the population is said to have this problem. Her research also indicates that infections interfere with Hg removal and her RNA based products Metals I, II, III and IV are designed to enhance immune response to herpes, measles and other viruses. After they are administered, Hg becomes easier to remove. Dr. Rowen reports that he has found her products useful.

Dr. Boyd Haley, and coworkers found that Alzheimer's disease patients also have Hg in their brains and increasing difficulty in eliminating it as the disease progresses. Dr. Rowen suggests that removing this metal from Alzheimer's and Parkinson's patients might also be beneficial in treating these diseases. The reluctance of the medical establishment to fund research on the connection between metals and diseases, mentioned by Dr. Haley in the past, may need to be overcome if these options are to be explored.

Dr. Bruce West's "Health Alert" letter of June 2005 provides some perspective to what he calls the autism epidemic. In 1996 1 child in 166 had autism and now a Japanese study indicates that it is affecting 1 in 62. This is only topped by the rate of diabetes, which was 1 in 16 people in 2002. For contrast 1 child in 154 had measles during the epidemic in 1925 and 1 person in 1,922 had cancer in 1992.

Dr. West asks: "Where are protective agencies"? & "Is the fox protecting the henhouse"? He also notes that Congressman Burton (R-IN) has said that the FDA and other agencies of the government are "asleep at the switch."

Considering the fact that the medical establishment, according to Daniel Haley 's book *Politics in Healing*, appears to ignore advancements in medical science, not of its own invention, it may be that it is just being consistent. However the elimination of thimerosol from animal vaccines in 1990 may have been the means for collecting data on animals to justify the FDA's request that this substance be eliminated from pediatric vaccines in 1999 which some say was not completely acceded to until 2004. But this is pure conjecture.

HG IN ALZHEIMER'S & PARKINSON'S DISEASES

Dr. Haley and coworkers at the University of Kentucky found that Hg was present in the brains of Alzheimer's patients and that there was evidence that they had increasing difficulty eliminating it as the disease progressed.

Dr. West notes in his June 2005 "Health Alert" that in 1977 Dr. Hugh Fudenberg, MD, a member of the WHO immunology panel found that older people who had 5 consecutive flu shots between 1970 and 1980 had a 10 times higher than normal rate of developing Alzheimer's disease due to the Hg and aluminum in the shots.

Dr. Rowen suggests that Dr. Yasko's products can benefit neurologically impaired adults, those with Azheimer's and Parkinson's diseases. They are said to eliminate infections, which impede removal of Hg. The fact that integrative physicians find repeated heavy metal chelation tests are negative when lots of mercury is present, as in the case of Dr. Buttar's son, clearly indicates that there are impediments to its removal.

A small scale test of the effect of Dr. Yasko's products and a chelation agent like DMPS on early stage Alzheimer's patients should not be very expensive and might open the way to a new therapy for this devastating disease.

If the removal of Hg from the brain can be found to improve the condition of Alzheimer's disease and Parkinson's disease patients, as Dr. Rowen suggests, this therapy will be hard to ignore. The potential benefits of the technology which the Autism Research Institute and its cooperating physicians and scientist have discovered, $billion reductions in healthcare costs, should be very tempting to politicians desperate for funds - and important to the voters who put them in office.

HG IN THE GENERAL POPULATION

According to a Wall street Journal article in 2004, 55% of Hg emissions come from natural sources: volcanoes and forest fires, for example. Nevertheless anthropogenic sources, like coal-fired power plants in China do create problems for us. According to the Center for Science and Public Policy about 70% of this country's Hg pollution is imported from Africa and Asia on the prevailing winds. Gold mining in Brazil is said to be a significant source of Hg entering the oceans. We can't do much about these things.

Our coal burning power plants are responsible for 1% out of the 3% of the world's Hg pollution produced in the US. We have made a lot of progress in curtailing Hg emissions and our industry appears to be only a minor threat to our well-being compared to foreign sources over which we have no control. Setting unrealistic goals for our power plants is unlikely to benefit either the People or the country.

There is good reason for believing that so-called silver fillings, which contain about 50% Hg, are an important factor in neurological diseases. The Dental Establishment maintains that they are not, but Dr. Haley can provide convincing evidence that they generate Hg vapor in people's mouths. He also has data indicating that a mother with 10 or more fillings in her mouth is likely to have an autistic child. Eliminating Hg from dentistry is a worthwhile and achievable goal.

Fish has recently been demonized as a source of Hg. Fish high on the food chain do contain a lot of Hg. But our bodies are designed to take care of a certain amount of heavy metals and most of the capability, about 66%, is found in the digestive system. Cold water fish, in particular, are good sources of omega-3 oils, a major deficiency in modern diets. Eliminating them would be a bad mistake.

The Center for Science and Public Policy claims that there is no epidemiological evidence that fish cause mental defects. Dr. Philippe Grandjean, MD of the Harvard School of Public Health claims that a Faroes Island study proves otherwise and that it was selected by the National Academy of Sciences as the "Gold Standard" for setting Hg exposure limits. But this study was of the effect of a diet of whale meat - not fish! Whale meat and blubber contain a rich soup of PCBs and other contaminants as well as Hg.

Furthermore the National Academy of Sciences imprimatur is not as convincing as it might seem when you consider that their report of an impending ice age in 1975 was quickly supplanted with one predicting global warming after notification that this was the unofficial policy of the United States. Politically modified science is not uncommon.

It is comforting to know that there was another study of the long-term consumption of a lot of fish by mothers in the Seychelles Islands, which concluded that there were no negative effects on their children. These mothers didn't consume whale meat, but 10 times as much of the same kind of fish as consumed in this country. The health of their children was followed from birth until they were 9 years old.

The threat of heavy metals damage can be countered with chelation therapy. EDTA chelation is approved by the FDA for lead removal. It also benefits cardiovascular disease, but is relatively ineffective in removing Hg. The agents mentioned in this article are not approved by the FDA for removal of Hg. The major impediment to use of these therapies is the Medical Establishment. Its suppression of chelation by harassment and persecution of providers is deplorable. Ten years ago 3 physicians provided this service in the Northland where I live. All were forced out of business. Possibly they were too much of a threat to the income of their brother physicians!

PROPOSED DMSA STUDY

Dr. James Adams, a chemical engineering professor at Arizona State University, who has an autistic child, has applied for institutional review board approval for a clinical evaluation of DMSA (dimercaptosuccinic acid) chelation as a treatment for autism. This is the first step in the process of establishing the validity of this therapy which has already benefited thousands of autistic children. The next step will require the cooperation of the FDA in the form of approval of a clinic trial.

Meanwhile Dr. Boyd Haley, chairman of the Department of Chemisty of the University of Kentucky is developing new chelating agents which, hopefully, will be even better at crossing the blood-brain barrier than DMSA. If Adams and Haley are successful, the treatment of autism will experience a quantum leap out of the psychiatric morass in which it has languished since it was first discovered, as an exclusively American disease, in 1941.

Presently autism is a label describing a communication and behavioral disorder of unknown cause. It has been the province of psychiatry for many years. However a review of the literature by Dr. S. Bernard et al. published in 2001 reported that the diagnostic criteria for autism were the same as those for mercury toxicity.

Biochemical research on the effects of mercury on neurons, enzymes and neurotransmitters by Dr. Boyd Haley and others during the last 15 years have established mechanisms which explain how the damage which results in autism occurs.

It is hoped that the authorities at Arizona State University and the FDA will move promptly to approve a clinic trial. This research proposal is based on sound scientific reasoning. The need for a rational treatment of autism is overwhelming.

There is no doubt that mercury is a developmental toxicant. CDC literature admits to this fact. Furthermore the CDC, EPA, and FDA have all established safe limits for the ingestion of mercury between 0.1 and 0.4 micrograms per kilogram per day. In addition the EPA recently reported over 300,000 infants are at risk for neurological damage every year because mothers in the U.S. have high levels of mercury in their bodies.

It has been found that children with autism have high levels of mercury in their bodies and only limited ability to excrete it. The severity of the disease correlates with mercury levels. Furthermore the victims generally improve when the mercury is removed.

DMSA, a derivative of succinic acid, is a chelating agent effective in removing mercury with positive benefits to autistic children. Unlike EDTA a chelator approved for the removal of heavy metals, DMSA is not blocked by the blood-brain barrier. Therefore it is able to remove mercury from that important organ.

Many physicians have successfully treated thousands of autistic children with this relatively benign substance. But it is not part of the "Standard of Care" which physicians use as a bulwark against the depredations of the legal profession.

This treatment is an important advancement in the state of the art. Those who were involved in its development deserve the gratitude of us all. It promises to remove burdens from families with autistic children, take a load off the elementary school system and curtail, to some extent, national costs of medical care.

It seems clear that we are all contaminated with varying

amounts of this dangerously neurotoxic substance. It enters our bodies with the air we breathe and the fish we eat. Mercury containing fillings in our teeth and mercury preservatives in vaccines add to the load. It is extremely likely that this substance could be a hidden cause of many diseases. It strikes at weak links in the central nervous system, for example the nerves that control the heart. High concentrations of mercury have been found in the hearts of athletes who have died suddenly and unexpectedly of idiopathic cardiomyopathy. It affects the brain. In my own experience it produced optical illusions, interfered with dreaming and caused delays in information processing. It has been found in the brains of Alzheimer's disease patients. It caused my teeth to loosen and an abnormally large number of cavities to appear.

Considering these facts, the existence of an approved therapy for removal of mercury could pave the way for advancements in medicine, which would have as objectives the cure of diseases rather than the treatment of symptoms.

Allegedly the Roman civilization declined and ultimately disappeared because of the lead derived from plumbing systems, which poisoned the Romans. The combination of mercury and lead, still found in the water supply of some cities, is even more dangerous.

COMMENTS OF NEW YORK TIMES ARTICLE ON AUTISM

The New York Times of June 25, 2005, in an article on autism, conveys the impression that parents with autistic children are uninformed and emotional with no scientific basis for their complaints about thimerosol in pediatric vaccines.

Dr. Melinda Wharton, deputy director of the National Immunization Program, suggests that this is an era wherein, it appears, science is not enough. It seems that Dr. Wharton may be uninformed about what competent scientists know and have discovered about autism and mercury toxicity. Or is she taking part in a cover-up for the protection of the medical establishment from the 4,800 suits that the Times says has been filed by parents of autistic children?

The Times provides full coverage for all efforts to discredit the concerned parents. For example it notes that Kristen Ehresmann from the Minnesota Department of Health proposes that mercury in autistic children is not coming from thimerosol in vaccines, but from fish, tap water or air.

There is no mention that children, on the day they were born, have been getting an injection of ethyl mercury that would only be safe, by EPA standard, if they weighed 275 pounds according to Dr. Boyd Haley, chairman of the Department of Chemistry at the University of Kentucky.

The statement that ethyl mercury is less toxic than the methyl mercury used in EPA safety tests is incorrect. This makes one wonder about the scientific competence of its sources of information.

It is incomprehensible that the "scientific" experts of the CDC, FDA, Institute of Medicine, WHO and the American Academy of Pediatrics can dismiss the notion that thimerosol causes or contributes to autism in the face of the scientific facts.

The facts are that autism first occurred in the United States as a new, rare disease that appeared at birth. It was still a rare disease in 1956 when Dr. Bernard Rimland's autistic child was born. That caused him to form the Autism Research Institute.

Now autism is no longer a rare disease. It affects one in 166 children, and it commonly shows up at 18 months.

Autism burgeoned after the National Immunization Program got under way with thimerosol-laced vaccines. Eliminating relatively benign childhood diseases has been very costly.

Nowhere does the *Times* mention that autistic children have great difficulty eliminating mercury from their bodies. This was established by Dr. Haley, Dr. Amy Holmes and Mark Blaxill, and it was confirmed by other research groups of scientists.

But the *Times* article goes to great lengths to warn about the dangers and costs of chelation for the removal of mercury from autistic children. The chairwoman of a panel from the Institute of Medicine, which reviewed this subject in 2004, was mentioned in the Times article, discussing potential harm, excessive cost and unproven technology to scare people away from this effective therapy.

The fact that chelation has apparently worked to improve the lot of thousands of autistic children apparently means nothing without a multimillion-dollar investment in double-blind studies, which the medical establishment is unwilling to fund.

It would be interesting to know what the safety investigations, which the CDC must have performed prior to the institution of the National Immunization Program, say about mercury and thimerosol.

CHAPTER 4
MEDICAL ESTABLISHMENT INTRANSIGENCE

There is no doubt that a majority of physicians are imbued with the desire to heal their patients as fellow human beings. However, their educational experience, with its heavy requirements of time and money, tends to replace altruism with cupidity. Along with the award of the coveted MD degree is a requirement to start paying on a heavy debt. This focuses the minds of the graduates on the economic aspects of medicine and has resulted in the production of a system based on specialization, which ultimately increases the total costs of medical care.

To insure that the public receives adequate and safe medical care, our laws provide for a national medical establishment. We have National Institutes of Health, the Federal Drug Administration, and other branches of the government, which are supposed to insure that the People receive the very best of medical care. Unfortunately, like all of our Governmental Institutions, they are subject to political and financial influences and, like all bureaucracies, they attempt to increase their power and influence.

For the protections of members of the profession and advancement of their influence, medical self-policing organizations have been formed. The first of these was the Homeopathic Medical Association, quickly followed by the American Medical Association. Early rivalry between these two was resolved by the discovery of antibiotics, which gave the American Medical Association and its allopathic physicians a well publicized advantage in the treatment of infectious diseases, the primary cause of death at the time. Chiropractors also formed an association and, in an attempt to reduce competition, the American Medical Association went too far and was convicted of restraint of trade.

Even before this, unofficial joint action by groups of physicians has been used very effectively to protect established practices of the medical community. Perhaps the earliest evidence of this was the Ignaz Semmelweis affair in Europe. This young physician, in charge of a charity maternity ward, began requiring attending physicians to wash their hands in chlorine water before treating their patients. As a result, the death rate in his ward dropped significantly. Instead of emulating this achievement, his peers persecuted him and consigned him to a mental hospital wherein he died. It is reported that Oliver Wendell Holmes, an American physician who became a Supreme Court Judge, also advocated the washing of hands between patients and was also vilified for his temerity.

As a result of my access to the deceased Christopher Bird's papers and library, I was able to acquire some familiarity with the many successful efforts by organized medicine to suppress advancements of medical science. Subsequently, I wrote about this in the following articles. When I visited Dr. Gaston Naessens in Canada as part of my research, I found that Daniel Haley, a former member of the New York State Legislature, had written a book, *Politics in Healing*, which confirmed my conclusions.

SUPPRESSED CANCER CURES

The medical establishments of several countries do not encourage the development of alternative treatments for cancer. In many cases, they even appear to be vindictive toward anyone, including physicians, who successfully treat cancer. A change in attitude of the medical establishment is long overdue.

"Despite many hundreds of clinical trials of chemotherapy during the past 40 years, breast cancer mortality has not decreased significantly while prostate cancer mortality has risen steadily and colon and rectum cancer remains high. Chemotherapy is not winning the war on cancer," said Dr. Myron Pollycove of the Nuclear Regulatory Agency at the 2004 Pacific Basin Nuclear Conference in Hawaii in March.

The main body of Dr. Pollycove's speech discussed 40 years of research that showed low-dose, whole-body radiation treatments stimulate immune and DNA repair systems of mice, rats and humans.

In 1976 and 1979 the results of clinical trials of this therapy at Harvard University Medical School were published. The 1976 trial resulted in 70 percent of 25 patients with non-Hodgkin's lymphomas survived four years. Only 40 percent of a matched group of 24 who did not get this treatment survived four years. In the 1979 trial, 74 percent of 39 patients receiving this treatment for the same disease survived, compared with 52 percent survival of 225 controls.

In 1997 Dr. Kenkichi Sakamoto and his co-workers at Tohoku University in Sendai, Japan, published the results of treating non-Hodgkin's lymphoma patients with 150 rads of whole body radiation at the rate of either 15 rads two times per week or 10 rads three times per week.

Eighty-four percent of those who received these treatments were still alive at the end of 13 years. Only 50 percent of the controls were alive after nine years. Previous to the low-dose therapy, all the patients had chemotherapy and localized high-dose tumor irradiation. Sakamoto also found that half-body irradiation (from hips to chin) was equally effective.

Some of the difference between the 1979 Harvard trial and the 1997 Tohoku trial may be due to the well-established benefits of lower caloric intake and more exercise of Japanese patients as compared to Americans.

When Sakamoto found that he had metastasized colon cancer, it is significant that, after surgery, he had the same low-dose whole-body radiation treatment that he gave his patients. The only difference was that he gave himself a second series. His metastases are reported to have disappeared, and he is said to be vigorous and in good health.

It is reasonable to ask why, almost 30 years after that promising Harvard trial in 1976, low-dose, whole-body radiation is not commonly offered to American cancer patients.

It is also reasonable to ask why medical establishments suppress promising alternative treatments for cancer when their own "approved" treatments have less than perfect results.

GASTON NAESSENS

A recent case is that of Gaston Naessens, a Frenchman who developed a microscope capable of over 20,000 times magnification, which he used to detect disease   processes in living blood. With it he discovered that healthy human blood contained what he called "somatids," tiny, living bodies, which, when a person's immune system deteriorated, would produce bacterial and "mycelular" (multi-celled) entities capable of damaging the body. He did not know that he was confirming work done by Dr. Antoine Bechamp recorded in his book Le Sang, published in 1901, a translation of which, titled The Blood, was published in 1912.

Naessens also developed immune-enhancing substances that reportedly helped cure large numbers of cancer patients in France. An investigation by the Surete is said to have found that these immune system enhancers benefited 65 percent of 10,000 patients. He was prosecuted for the illegal practice of medicine and fined 18,000 francs and costs, along with an extra 5,000 francs awarded to the Doctors Association of the Seine Department.

Being warned that another indictment was planned, Naessens fled to Canada. He settled in Quebec, where a foundation provided funds for a laboratory. He again began research on immune enhancers, which resulted in a patent for a camphor derivative called 714-X. Cancer and AIDS patients flocked to his door.

But once again he was charged with the illegal practice of medicine and fined and then he was charged with murder because a terminal cancer patient who had taken 714-X had died.

Then Naessens's luck changed. Christopher Bird, author of a number of books, including *The Secret Life of Plants*, became aware of his problems, helped raise funds for his defense, monitored the trial and ultimately wrote a book about it. Naessens was also fortunate in having some prominent witnesses testify on his behalf, including a French ambassador whose wife was benefited by Naessens's products. A jury declared him not guilty. However, a Dr. Roy, head of the local medical society, was not pleased and made disparaging remarks about the jury.

DR. ROYAL RAYMOND RIFE

Another genius with a microscope and cancer treatment met a far less happy fate at the hands of the American medical establishment.

In 1944 an article describing two microscopes, the new RCA electron microscope and Royal Raymond Rife's "universal microscope," was published in both the Smithsonian Institution's annual report and The Journal of the Franklin Institute. Most of the article was devoted to a description of the Rife microscope, which permitted viewing living matter at over 60,000 times magnification with excellent resolution.

Rife was the first person to see a virus in living tissue. He found that bacteria and viruses could be made to fluoresce at specific wavelengths of light and that many could be identified by the color of their fluorescence. He also found that there seemed to be frequencies of radiation that could destroy them.

Subsequently Rife developed an electromagnetic device for killing bacteria and viruses, which was reported to be effective against cancer.

This and his microscope created a furor. The American medical establishment soon had him in court. His electromagnetic device was declared illegal. Rife's trial caused him to have a nervous breakdown, followed by alcoholism. He is reported to have died as the result of administration of an inappropriate drug while he was recovering from a binge in a hospital.

A search for the five microscopes that Rife built resulted in the conclusion that one of them, which is inaccessible in storage in England, might be usable.

DR. WILHELM REICH

Dr. Wilhelm Reich used another unusual microscope. He is said to have been Sigmund Freud's star pupil. Unpopular with the American Food and Drug Administration, his books and papers were burned in an incinerator in New York City, and Reich died in jail in 1976.

Reich worked with two microscopes with 150 times objectives and 25 times oculars. One had additional magnification built into the viewing assembly, which enabled him to reach 5,625 times magnification. While these very high magnifications did not produce high resolution, they did allow for the detection of movement. With them he found tiny elements, which he called "bions," in living blood. He also found what he called "T bacilli" in the blood of cancer patients, which were not present in the blood of healthy people. The "T" stood for "Tod" or "death" in German.

Reich also discovered a new form of energy, which he called "orgone," and he developed equipment for studying its properties. After experimenting with mice, he built larger equipments called Orgone Accumulators for experiments with people.

It is reported that they were useful in treating diseases, including cancer. They also had a dangerous capability of enhancing radioactivity. An experiment with a small amount of radium had bad, but fortunately not fatal, effects on every member of his family.

Reich, unfortunately, had radical political and social ideas. A freelance newspaper reporter named Mildred Edie Brady interviewed him and wrote an article in 1947 headlined: "The New Cult of Sex and Anarchy—The Strange Case of Wilhelm Reich." Inaccurate and slanderous reviews followed.

Dr. Austin Smythe classified the Orgone Accumulator as a fraudulent cancer cure in *The Journal of the American Medical Association* and stated that there was no evidence to support Reich's claims. This was reprinted in *Consumer Reports*. The president of the American Psychiatric Association said Orgone was a fake. Dr. Potter told a physician using an Orgone Accumulator that the APA, the AMA and the FDA were exerting efforts to put a stop to it once and for all.

On the other hand, physicians, who had confirmed his clinical claims, formed the American Association for Medical Orgonomy. They were reported to have had many detailed case studies supporting Reich's findings.

Ultimately the FDA charged Reich with fraud on the basis that there was no such thing as orgone and therefore his Orgone Accumulator didn't work. Reich soon found himself in court, was convicted and was sent to prison. His equipment was destroyed.

However, it is reported that several physicians who were using the accumulators asked the judge who had presided at the trial whether they were affected by the court's decision. The judge held that, since they were not named in the indictment, they were free to use their equipment.

DR. VIRGINIA LIVINGSTON

In 1970, Drs. Virginia Livingston and Eleanor Jackson published a full description of progenitor cryptocides, a tiny bacterium found in tumor tissues and body fluids of cancer patients, which passed through filters designed to separate bacteria from viruses.

In 1976, Livingston was granted a patent, No. 3958025, for the treatment of a vitamin deficiency of abscisic acid in man, animals and avian species as related to the prevention or suppression of cancer by this vitamin.

A clinician, Livingston is reported to have treated many cancer patients, including Dr. Owen Wheeler, her future husband. Together they wrote five books on the microbiology of cancer.

Shortly before her death she was ordered by California medical authorities to cease treating cancer patients because her method of treatment was not in accordance with the law, which stipulated exactly how cancer patients were to be treated.

DR. STANISLAW BURZINSKI

More recently Dr. Stanislaw Burzinski developed and patented "antineoplastons," which proved effective in curing some, but not all, kinds of cancers by causing the cancer cells to revert back to normal. Although engaged in FDA-approved trials, his Texas office was raided by armed FDA agents who removed his patients' records. Were it not for the protests of his patients, who were successful in obtaining a congressional hearing for him, he would have been put out of business and possibly jailed.

RENE CAISSE

The case of Rene Caisse, a Canadian nurse, is interesting. She was able to help many cancer patients with "essiac," an herbal preparation said to have been used by native American Indians. Ultimately she was allowed to continue providing her services to Canadian patients as long as she was willing to forego receiving any remuneration from them.

WASTEFUL RESEARCH

In 1987 Rep. Henry Waxman (D-Calif.) introduced a long statement by Dr. Samuel Epstein of the University of Illinois Medical Center into the Sept. 9, 1987, issue of *The Congressional Record*. Among other things, Epstein said:

> Congress has tended to abdicate decision making to scientific authority (or perceived authority), rather than questioning its basis in the open political arena. Of particular importance was passage of the 1971 Cancer Act in response to orchestrated pressures from the "cancer establishment," the National Cancer Institute (NCI), American Cancer Society and clinicians pushing chemotherapy as a primary cancer treatment.
>
> The cancer establishment misled Congress into the

unfounded and simplistic view that the cure for cancer was just around the corner, provided that Congress made available massive funding for cancer treatment research. The Act did just this, while failing to emphasize needs for cancer prevention, and also gave the NCI virtual autonomy from the parent National Institutes of Health, while establishing a direct chain of command between the NCI and the White House.

Some 16 years and billions of dollars later, Congress still has not yet appreciated that the poorly informed special interests of the cancer establishment have minimized the importance of and failed to adequately support critically needed cancer prevention efforts. Nor has Congress appreciated the long overdue need for oversight on the conduct and priorities of the NCI. Given the heterogeneity of congressional interests, the complexity of the problem involved the heavy industry lobbying, the indifference of the general scientific community and the well orchestrated pressure of the cancer establishment, it is not surprising that Congress has still to recognize that we are losing the war against cancer.

Waxman's staff told Whole Body Health that, today, he does not necessarily agree with the above statement but, in his opening statement as ranking member of the Committee on Government reform, on May 13, 2004, said: "Over the last several decades, while rates of heart disease have dropped dramatically, rates of cancer have largely remained stable. Despite progress against a few specific tumors, cancer will still kill an estimated 500,000 Americans in 2004."

LOSING THE WAR

If we were losing the war against cancer in 1987 and we are still not winning it 17 years and millions of dollars later in 2004, isn't it time to stop following a policy that very clearly has failed?

There are alternatives for which the medical establishments have successfully prevented a fair trial. Rife's microscope and electromagnetic device and Naessen's microscope and immune enhancers, for example, would appear to have considerable merit.

However, it may be necessary to reduce the power of the American Medical Association, which was convicted of anti-trust activities in the past, as well as the Food and Drug Administration and other members of the medical establishment, in order to give these alternatives to the standards of care a chance.

Unfortunately Reich and Rife are dead. But Naessens is still alive, although over 80 years old. Naessens uses a microscope of his own design which has been designated an advancement in the state of the art and has developed a condenser that permits users of other kinds of microscopes to observe what he can see with his instrument. It would be a real shame to have these disappear with his death.

Naessen's immune enhancer has been used to treat over 4,000 Canadian patients via prescriptions by 1,500 Canadian physicians.

In the United States, Charles Pixley, who for a time was distributing this product, was prosecuted and jailed by the FDA in the 1990s.

Initiating research on enhancement of the human immune system, which was profitable for Naessens, might produce new treatments for cancer and other diseases more benign than chemotherapy, and less expensive, too.

CHELATION THERAPY

Over 20 years ago in Switzerland, scientists from the Institute for Radiation Therapy and Nuclear Medicine at the University of Zurich discovered some interesting and important facts. The first was that people living in high traffic areas were more likely to develop cancer than those who lived in traffic-free areas. This one might expect, since lead from tetraethyl lead, used to increase the octane number of gasoline at that time, is definitely toxic. The other fact was unexpected: Of the 231 people who lived in high-traffic areas in their sample, 59 appeared to be protected against cancer despite the exposure.

Aware of the importance of this discovery, and also of the controversy that might arise concerning it, Drs. W. Blumer and T. Reich submitted their data to a skeptical epidemiologist at the university for review. He could find nothing wrong, so they published their findings in *Environmental International* in 1980.

Their paper disclosed that only 1 of the 59 people, who had received chelation therapy, died of cancer during their 18 years of monitoring of the group exposed to high traffic. On the other hand, 30 of the 172 people who did not receive this treatment—the rest of the 231—died of cancer during this same period.

In other words, chelation therapy reduced the cancer rate in the protected group by 90 percent. Deaths from all causes were also lower in the chelated group.

Dr. Bruce Halstead, who wrote *The Scientific Basis of Chelation Therapy*, considered the reduction of dangerous levels of free radicals, resulting from removal of iron, copper and the heavy metals from the bloodstream, to be responsible for the major benefits of this therapy.

Dr. H.B. Demopoulos of New York Medical Center found that heavy metals, free radicals and oxidized fats were associated with the initiation of cancer.

Consideration of these facts and opinions provides a plausible basis for the conclusion that chelation therapy may provide a means for removal of causes of diseases.

This should be far more cost-effective than treating the symptoms of diseases. It just might be the means for blunting the continual escalation of medical expenses, which now amount to about 12 percent of our gross national product.

Dr. Elmer Cranton's *Bypassing Bypass and Forty Something Forever* by Harold and Arlene Brecher contain a great deal of information on this therapy and should be readily accessible to the average citizen.

EDTA, also known as edetic acid, but more commonly known as ethylenediaminetetraacetic acid, is a chelating agent. This means it is capable of taking metal ions from solution and sequestering and concealing them inside its structure, thus making them incapable of their usual activity. Chelating agents make hard water soft by hiding calcium and magnesium. They take metal ions out of the bloodstream and sneak them out through the kidneys during chelation therapy.

EDTA's use for removal of heavy metals from the bloodstream was approved by the FDA many years ago as a safe and effective treatment.

About 10 years ago, Dr. Ross Gordon, former president of American College for Advancement in Medicine (ACAM) stated that almost half-a-million people—some 20 percent of them given up for dead—had received this treatment.

Overall about 85 percent had benefited. There had been fewer than 30 deaths. Most of these had occurred before the development of protocols during the early research. Adverse reactions were rare.

Despite this, the medical establishment has attempted to dispute these findings, going so far as to try to prosecute practitioners of chelation therapy.

According to The Scientific Basis of Chelation Therapy, the AMA considered chelation therapy for the treatment of arteriosclerosis and related disorders as useless "because the effects are not long lasting." This opinion was issued in 1966.

In 1976, the California Medical Association Council echoed the AMA's opinion, saying: "the efficacy of chelation therapy in the treatment of arteriosclerosis is not proven and is not now an accepted therapy."

The following year, the California Department of Health and the Board of Medical Quality Assurance attempted to prevent physicians from using EDTA for chelation by threatening criminal prosecution. They were frustrated, however, by the California attorney general.

In 1978 the FDA attempted to obtain an injunction against Dr. H.R. Evers who was giving chelation treatments in Alabama. In the case brought against Evers, FDA officials argued that there was a strong medical school of thought that chelation had not been clinically proven. But the weight of evidence submitted was to the contrary, and the court ruled against the FDA.

In another case, a Federal Trade Commission judge determined that the AMA had produced a formidable impediment to competition in the delivery of health care services to physicians in this country.

"That barrier has served to deprive consumers of the free flow of information about the availability of health care services, to deter the offering of innovative forms of healthcare and to stifle the rise of almost every type of health care delivery that could potentially pose a threat to the income of fee-for-service physicians in private practice," said the court.

Daniel Haley, in his remarkably informative book, *Politics in Healing*, notes that the Berkeley, Harvard and Mayo Clinic newsletters condemned chelation therapy in 2000 presumably because it was unproven.

To insist that a procedure is unproven, which has benefited millions of people with fewer than 30 deaths reported, flies in the face of reason. Especially since the U.S. Office of Technology Assessment, a congressional watchdog agency, has determined that only 10 to 20 percent of procedures currently used in medical practice have been shown to be efficacious by controlled trials.

In 1998, the FTC took ACAM to court based on a brochure's statement that chelation is an effective treatment for CVD. Since ACAM did not have two double-blind, placebo-controlled studies to support this claim, the FTC asked the court to place ACAM on probation for 20 years.

It remains a mystery why the FTC contradicted its earlier

finding in the Evers decision in 1978 and went after ACAM.

Why are the bureaucrats in the FDA and medicrats in the AMA and state boards so intent on eliminating chelation therapy? Might it have something to do with the difference in cost of EDTA treatments for cardiovascular disease—about $3,000—as compared with heart surgery, which can run as much as $50,000?

Haley believes that people need to get "pitchfork mad" and insist on freedom of choice in medical care.

Haley's recent experience with a stroke illustrates his point that this freedom does not now exist. Haley's health care providers refused to give him a hyperbaric oxygen treatment or an infusion of vitamin C to help his recovery.

EDTA preferentially removes heavy metals like lead. It is least effective in removing magnesium.

Early on, the removal of calcium "rivets" holding plaque to the walls of arteries was believed to be the mechanism by which chelation helped people with CVD. The mechanism is now known to be far more sophisticated.

Less than one-half of a gram of ionic calcium removed from the bloodstream during chelation treatment lowers its concentration and triggers a response from the parathyroid gland.

The gland sends out its messenger, a parahormone, which tells body systems to scavenge calcium from storage to replace the amount removed.

Also, providentially, this hormone stimulates the bones to put more bone building into operation to protect the skeleton. The bone-building effect is said to continue for three months and results in stronger bones.

The human skeleton is said to contain between 1,000 and 1,500 grams of calcium. The skeleton is a dynamic structure existing as highly insoluble apatite, a calcium phosphate mineral. The skeleton continually adjusts to the changes in stress encountered during a lifetime.

Plaque forms as a result of damage, either caused mechanically or due to attack by free radicals. The process starts with the formation of an atheroma, a little tumor, with no blood supply of its own. As it grows, its interior deteriorates, because diffusion of oxygen and nutrients through the thickening walls becomes inadequate. In the final stages the deteriorated regions fill with cholesterol and calcium. Furthermore, oxidized cholesterol acts like vitamin D and assists in the hardening process.

So how does chelation get rid of the plaque? The simple, perhaps incredible, answer seems to be that it clears out the heavy metals and copper and iron, which poison the body. The body then can heal itself. That is, if the proper nutrients are supplied in adequate quantities and care is taken to minimize the introduction of more of the poison that caused the problem.

Furthermore, since these same enzyme poisons are responsible for a lot of other malfunctions, this one treatment can logically be expected to have manifold benefits. And it does.

It should be noted that chelation creates problems for at least two influential groups of people: The health insurance companies who like to pay for single treatments for specific diseases and fee-for service physicians who like to treat conditions one at a time with a specific procedure for each one.

Health insurance companies generally will not pay for chelation treatments unless forced to by the court. The medical establishment, through the FDA, the AMA and state medical boards and many physicians in private practice, has been engaging in harassment of those who provide chelation for years—and is still doing it.

GHARDIALI, RIFE & BURZINSKI'S GENTLER THERAPIES

New research shows promising results for a number of alternative therapies as a way to treat people suffering from a broad range of illnesses. The May 2003 issue of *Scientific American* reports that NASA-designed red light emitting diodes were found by astronauts in orbit to not only encourage plant growth but also heal wounds.

It is suggested that this colored light stimulates the mitochondria in human body cells, by some unknown mechanism. Dr. Janis Eels at the Medical College of Wisconsin also reported that red LED light pulses enabled methanol-poisoned rats, which normally are blinded, to recover 60 to 70 percent of retinal function.

In addition studies of the effects of laser light have been under way for several years.

The use of colored light for healing purposes has a long history. The ancient Egyptians and the ancient Greeks are reported to have used this modality. In more recent times a British physician wrote a book on this subject.

Building on his research, Dr. Dinshah Ghardiali, a Persian who came to this country from India, developed a complete system of healing for use with devices called the Itsitometer, for local body temperature measurements, and Spectrochrome for irradiating areas of the body with selected colors of visible light. Dr. Kate W. Baldwin, who had been senior surgeon at the Philadelphia women's hospital for many years, reported to a clinical meeting of the eye, ear, nose and throat diseases section of the Pennsylvania Medical Society on Oct. 12, 1926, that colored light is the simplest and most accurate therapeutic measure yet developed. She said that she had been experimenting with colored light therapy for six years and that, after 27 years of active hospital and private practice, she believed that she could produce quicker and more accurate results with colors than with any and all methods combined and with less strain on the patient. She used a number of Spectrochrome devices in her practice. To illustrate the point, she mentioned the case of an 8-year-old child, with extensive burns, brought to her with a temperature over 105 degrees. According to Baldwin, there had been almost complete suppression of urine for over 48 hours, and the child did not respond to forced fluids.

Baldwin irradiated the kidney area with scarlet light (from the Spectrochrome instrument) for 20 minutes. Two hours later the child voided eight ounces of urine. The burns healed with less pain and scarring than would normally be expected. Baldwin suggested that the use of color in the treatment of burns was well worth investigating by every member of her profession. Baldwin also mentioned the case of a woman, with a carbuncle involving the back of her neck from mastoid to mastoid and from the occipital ridge to the first dorsal vertebra. The woman came to her for color therapy after 10 days of the very best of orthodox medical care.

After the application of colored light, no opiates, or even sedatives, were required. The patient was spared much suffering and scarring.

She also reported success with septic conditions, cardiac lesions, asthma, hay fever, pneumonia, corneal ulcers, glaucoma and cataracts to illustrate the wide applicability of color therapy.

One would expect that Ghadnali would be rewarded for spending 30 years developing the instrument, whose use Baldwin described, and for educating physicians and others about this technological advancement. However, records indicate that he was instead forced to defend himself in court numerous times, bankrupted and, finally, jailed. His books, other than his personal library, were ordered to be burned by the FDA. His instruments were declared illegal and destroyed.

Another medical innovator whose remarkable achievements were met with disapproval was Dr. Royal Rife. He designed and built an unusual microscope said to have had a magnification of 35,000 times with which he was able to see a virus in its natural state.

He found that bacteria and viruses were susceptible to destruction by radiation from an argon lamp fluctuating at frequencies in the sonic range. His microscope was wrecked, his machines declared illegal for use and destroyed; and, his reputation ruined. He became an alcoholic. Subsequently, he died in a hospital after being given, according to reports, a drug inappropriate for alcoholics.

Dr. Stanislaw Burzinski's case is a more recent example of an attempt by the FDA to suppress medical innovation. A brilliant physician trained in Poland, Burzinski came to this country and developed a cancer treatment based on antineoplastons. These are able, in many cases, to cause cancerous cells to revert to normal.

Although he was curing otherwise helpless brain cancer patients and engaging in FDA approved trials of his therapy, FDA personnel raided his clinic and confiscated his patients' files in an attempt to put him out of business. Public outcry by his patients and their families coupled with a congressional hearing forced the FDA to back off in this case.

Jerry A Green, J.D., at the Light Years Ahead Conference held in 1992, noted that the difference between the traditional medical practitioner and the wholistic practitioner is based on more than 2,000 years of divergent methods of theoretical understanding and scientific inquiry. The current allopathic model became dominant after the discovery of antibiotics. These enabled the allopaths to displace homeopaths from preeminence in the United States.

Medicine as practiced by allopaths is a dangerous business.

Reliance is based on surgery, hard radiation and drugs. Medications are generally poisonous substances dispensed in less-than-lethal doses. A few years ago The New England Journal of Medicine reported that prescription drugs kill about 180,000 Americans a year. Surgical mistakes and overexposure to radiation make a currently unknown contribution to the death rate. Modern medicine has had great success in the treatment of traumatic injuries and infectious diseases, although overuse of antibiotics appears to be generating a new breed of hard-to-kill bacteria. However, ever-increasing costs of medical care, especially the costs of medications, are meeting budget constraints and forcing changes in the practice of medicine.

Therapies such as those made possible by the Spectrochrome and Rife machines would appear to be both less costly and less lethal than drugs and radiation. Perhaps it is time for them to be revitalized and re-evaluated. And, just possibly, posthumous honors might be bestowed on the inventors of these technologies who were so badly treated while they were alive.

PLEOMORPHIC MICROBES IGNORED

The idea that cancer is a local disease that spreads, or metastasizes, is widely accepted. On the other hand there is a large body of research, which clearly indicates that systemic diseases, initiated by deteriorating conditions inside our bodies, are possible, and cancer might be one of them. If this is correct then many of the currently popular cancer treatments need to be reexamined.

Over 100 years ago in France, Antoine Bechamp, a biochemist and dean of the medical faculty at the University of Lille, found that tiny microorganisms, which he called microzymas, were present in the blood and other tissues of the body. He also found that they could change from one form into another. This is called pleomorphism.

At the same time, in Germany, Robert Koch and his students proved that certain bacteria caused tuberculosis and others caused other diseases like cholera. The identification of bacteria that caused particular diseases led to the hypothesis that bacteria were monomorphic. In other words that they did not change there form. The work of Louis Pasteur in France led him to agree with this hypothesis and soon it became an accepted theory.

Despite Bechamp objections, and much controversy between him and Pasteur, the theory of monomorphism was firmly established. Ultimately the world's medical establishments disregarded Bechamp's findings. Today, even in the face of extensive research by a number of scientists demonstrating that some microorganisms can change from one form to another, monomorphism is still the accepted dogma

Bechamp and his work are almost unknown now, although, his obituary took up about 8 pages in a technical journal after his death. Also disregarded by mainstream medicine are the extensive researches of Dr. Guenther Enderlein, a German zoologist who continued Bechanp's research, but on what he called protits instead of microzymas. According to Enderlein, a protit is a protein particle capable of developing into higher forms. It is alive, plant-like and apathogenic.

In the case of Mucor racemosus Fresen, Enderlein found a life cycle with 14 different phases including bacteria, fungi and spores. He found that several diseases were associated with different forms of this substance. Furthermore he developed biological preparations, which were found to effectively cure diseases by inactivating the pathogenic forms of Mucor racemosus.

Sixty years of research led him to conclude that (1) protits live within cells and are the primary living units in our bodies; (2) The blood is not sterile, but contains microorganisms capable of causing disease: (3) microorganisms undergo a growth cycle. His book *Bakterien Cyclogeny* was published in 1925 and gained him international recognition and honors. In 1931 Dr. S. A. Petroff, director of the Trudeau Sanitarium Laboratory in Saranac Lake, New York wrote "practically all American bacteriologists look upon your work as one of the most important contributions since Pasteur's and Koch's discoveries"

Two Americans, Drs. F Leonis and N. E Smith, published *Studies on the Life Cycle of Bacteria* based on research supporting Enderlein's theories.

Simply put, Enderlein found that in early stages of development a limited number of specific organisms lived in harmony with the body's cells and performed useful functions. But if the internal environment of the body deteriorated, these microbes changed, and in the process, became increasingly pathogenic and caused diseases.

It is said that Enderleine's ideas were not accepted by organized medicine because he was 100 years ahead of his time. He predicted that in the 21st century researchers would probably be studying a new world of biology involving objects much smaller than viruses. We are now in the 21st century. People are sicker than they should be and medical costs are becoming unbearable. Maybe it is time for the medical establishment to take a new look at what Enderleine found in the 20th. For this they will need to use more powerful microscopes. Those in current use in medical fields are limited to about 2,000 times magnifications. You can't even see viruses with these. Viruses have been observed with electron microscopes, but these are not suitable for studying the living processes that Enderlein discovered.

Dr. Erik Enby, a Swedish physician, who wrote *Hidden Killers*, said that nobody could heal cancer with any therapy. Only the body itself can heal. He asserts that the biological preparations discovered by Enderline help the body to heal itself. He claims that microbial activity, which can be observed in body fluids, supports the growth of malignant tumors. Depriving these tumors of their microbe supply with Enderlein's preparations causes them to collapse. Change these dangerous microbes into harmless forms and tumors stop growing. Cancers turn into scar tissue and patients get healthier

Dr. Windstrosser, a research associate of Enderlein, stated that it is not necessary to kill cancer cells, healing by a repair process is more beneficial for patients. He believed that chemotherapy and radiation could do more damage than cancer cells.

He claimed that cancer is a multi-causal illness and therapy needs to deal with all of the causes. For example blood pH of 7.2 to 7.6 reduces cancer risk, but changes in food supply and environmental pollution have increased people's blood pH. Nutrition is very important but many nutritionists seem to have been politicized. For example Linus Pauling's research on vitamin C is still not fully appreciated. Furthermore some physicians are still telling their patients that all they need is a balanced diet – supplements are unnecessary!

Dr. Wilhelm Reich one of Sigmund Freud's star pupils branched out into high magnification, but low definition, microscopy and discovered a tiny microorganism, which he called

"Tod" (death) bacteria in the blood of cancer patients. He also discovered a new form of energy, which he called "Orgone," that he and others found helpful to cancer patients. The FDA maintained that this Orgone energy didn't exist and destroyed his equipment. Dr. Reich died in jail soon after the FDA confiscated his books and papers and had them burned in an incinerator in New York City.

Dr. Elizabeth Livingston isolated and, subsequently with an associate, characterized a tiny microorganism, which she found in cancerous tissues and named Progenitor Cryptocides. It was so small that it passed through a filter, which was designed to separate bacteria from viruses. Nevertheless it could be cultured like a bacterium.

She found that this microbe produced a hormone and obtained a patent covering a method of controlling it. Towards the end of her career, after she married a Dr. Wheeler whom she had cured of cancer, she was ordered to stop using her therapy because it was not consistent with a California law, which specified how cancer patients were to be treated.

She and her husband knew microbes associated with cancer contained calcium. They also found that chelation with EDTA had removed them from the blood of one of their patients. This may help to explain why chelation therapy was found, by Swiss scientists, to be protective against cancer.

Royal Raymond Rife developed the "universal microscope" capable of over 60,000 times magnification and discovered what he called BX, a tiny microorganism that caused cancer in animals. He reported observing some microorganisms change into others when very small changes were made in their environments. He found that microorganisms could be made to fluoresce at specific frequencies, identifying themselves without staining

Rife also developed an electromagnetic device designed to destroy the BX, cancer causing agent and other disease agents as well. A trial involving 16 terminal cancer patients resulted in 14 of em, being completely cured in 3 months, most after 3-minute exposures at 3-day intervals.

His ray machines were used successfully by a number of physicians. Unfortunately they were confiscated and destroyed as illegal devices by the FDA. Several attempts have been made to duplicate his technology, but its illegal status limits economic incentives.

Rife himself was harassed through our legal system and became an alcoholic. A mysterious fire destroyed his laboratory records, including movie film demonstrating pleomorphism in bacteria. Not one of his universal microscopes has been located in operating condition in the U.S. One may be in storage in England.

Rife's microscope made the discovery of Enderlein's predicted world of biological objects smaller than viruses possible. In fact Rife discovered that there was such a world.

In the 20th century.

Gaston Naessens, a Frenchman now living in Canada, developed another high-powered microscope with outstanding resolution as a young man. Its performance puts it beyond the state of the art of commercially available instruments. It operates on different principles than Rife's microscope, but provides the same promise. With it he identified a 16 elements life cycle of a tiny microorganism, which he calls a "somatid" in the blood of patients, afflicted with degenerative diseases like cancer and AIDS. The somatid engages in a 3-element cycle in the blood of healthy people.

Unfortunately the medical establishments in Canada, France and the United States have not been willing to adopt the advances in medical science, which Gaston Naessens and those who preceded him achieved. Prosecution in France caused him to leave his native country. He was not welcomed in Canada and, in fact, the medical establishment there charged him with murder. Christopher Bird, author of the *Secret Life of Plants* and other books audited the trial. His book *The Persecution and Trial of Gaston Naessens*" provides extensive information about Naessens. An appendix contains information about Rife and his microscope.

It seems clear that medical establishments in several countries are not only disinterested in using advances in medical science, but actively discourage the adoption of those which might result in lowering the cost of medical care or more efficiently cure disease.

To allow remarkable instruments like the Rife microscope and Rife Ray machine to be destroyed could logically be considered an offense against humanity. It is to be hoped that the same fate will not befall Naessen's microscope.

For more than 100 years competent scientists have been finding tiny microbes like microzymas, progenitor cryptocides, Tod bacteria, BX and somatids and developing new therapies based on their findings. Medical establishments have been consistently ignoring these advances in medical science. They have not only ignored inventors of new and useful scientific instruments like the Rife universal microscope and ray machine and the Naessens microscope, but they have persecuted them as Christopher Bird has emphasized. How long must we put up with this intransigence which is certainly not benefiting our people. If, as is claimed, our Government is operated for the benefit of the people, Congress and the President need to take note of this unsatisfactory state of affairs. The power of medical monopolies, which they have been abusing, needs to be curtailed.

It seems to me that the poor results that have been derived from $billions spent on research during the war on cancer warrant a change in direction. Perhaps it would be smart to examine the basic assumption that cancer is a local disease that can spread. The evidence supporting the idea that it is a systemic disease seems to me to be persuasive.

POSSIBLE SUCCESSORS TO ANTIBIOTICS

It is common knowledge that antibiotics, the wonder drugs of the last half of the 19th century, are losing their luster. More and more bacteria are adjusting to their threat. This situation merits a look at alternatives, which might be gentler and cheaper than our presently available therapies—and will not lead to a whole new breed of super bacteria.

Mainstream medicine has inadvertently created new varieties that are very difficult and sometimes impossible to defeat. Medical statistics indicate that degenerative diseases are affecting more and more people. Some diseases are making a comeback, like tuberculosis, and the cost of medical care seems to have no upper limit.

Fortunately there is a whole century of research on biological preparations. They are not presently a part of the "standards of care" that guide and limit medical treatment in this country. But it should not be excessively difficult to adapt what has been developed and successfully applied in other countries to the needs of the American people.

Past history clearly indicates that there is a knowledge base for useful therapies for problem diseases, which is not being exploited by official medicine in America.

Many years ago Dr. Guenther Enderlein and others in Germany, using dark field microscopy, were able to relate elements of a growth cycle of a fungus to a number of chronic diseases. Medical establishments have ignored his research because they are unwilling to recognize that some microbes are pleomorphic, i.e., they can change their appearance and function.

In the 1940s Royal Raymond Rife developed a unique universal microscope with magnifications 60,000 times and found that small changes in microbes' environment could cause them to change from one kind of bacteria into another.

In more recent times, Dr. Gaston Naessens, a French-born scientist in Canada, using a microscope of his own design with magnifications up to 25,000 times with 150 micron resolution, identified another cycle that also relates to a number of chronic diseases.

Medical establishments have been unwilling to recognize the evidence that bacteria can change from one into another since the time of Pasteur. Consequently they have ignored these results as well as the biological preparations, which Enderlein and others developed for the alleviation and cure of chronic diseases.

Despite the disbelief of conventional medicine, remarkable progress has been made by independent physicians using this scientifically based technology to efficiently alleviate and frequently cure chronic diseases.

The following examples illustrate this point.

TUBERCULOSIS

In 1903 Prof. Friedrich Franz Friedman, M.D., a young Jewish physician in Germany, developed a biological medication for the treatment of tuberculosis, which reduced and sometimes eliminated the need for costly operations or long cures in sanitaria.

Operators of German tuberculosis sanitaria were not pleased and criticized his treatment vociferously. It was also a threat to the chemotherapy industry.

Nevertheless, by 1912 successful treatment of nearly 15,000 patients established its efficacy, and in 1914 Paul Ehrlich, a Nobel laureate, was put in charge of a commission to evaluate Friedman's discovery. In 1922 the Prussian National Assembly commission declared his medication harmless and surprisingly effective, if used early, against all forms of tuberculosis.

Enderlein studied the effects of Friedman's medication and found that one or two injections of it halved tuberculosis infestations in patient's lungs. The effect continued to destroy the disease until, in many cases, it completely disappeared.

Friedman lectured the U.S. Senate, by invitation, about his discovery. Subsequently the U.S. Government Printing Office is said to have published a 54-page document titled *Dr. Friedman's New Treatment for Tuberculosis*. President Teddy Roosevelt sent him a letter of appreciation wishing him success in his work.

Unfortunately an attempt to introduce Friedman's methods into the United States ran into a brick wall in Saranac Lake, N.Y., where the Trudeau Sanitarium for Treatment of Tuberculosis, and a thriving "cure cottage" industry was well established.

According to Alfred Donaldson's *History of the Adirondacks*, Friedman sold the "bottling rights of his cure" to an American syndicate that leased the Algonquin Hotel on Lower Saranac Lake for use as a sanitarium for the Friedman treatment of tuberculosis in the spring of 1913. The venture proved to be ill advised and ephemeral. The company closed its doors on Jan. 21, 1914.

What happened to Friedman's treatment demonstrates how mainstream doctors work to undermine treatments that threaten their existence and their profits.

Newspapers of that time reported that Dr. Edward Livingston Trudeau, the head of the Adirondack Cottage Sanitarium, had warned patients not to come to Saranac Lake for the Friedman treatment until it had been evaluated. Then he and his associates provided a local resident businessman with letters of introduction to colleagues in Berlin and sent him off to Germany to see what he could find.

Trudeau's colleague visited Friedman as a prospective patient, but refused treatment. He also contacted some of Friedman's enemies.

Returning to Saranac Lake he gave Trudeau two vials of bacteria obtained from a biochemist who had helped Friedman isolate his starting non-virulent tuberculosis culture obtained from a turtle in the Berlin Aquarium.

He said he believed that this was Friedman's vaccine, although Friedman was known to disagree. Trudeau reported that it was a live tuberculosis virus, nothing new, and may have been derived from material previously shipped from Saranac Lake to Berlin.

A wordy warfare raged around Friedman's name and his claims, according to Donaldson. As a result, the American people did not benefit from his advancements of medical science despite the efforts of Congress and plaudits of the president.

A 1915 newspaper report disclosed that Trudeau supplied

background information to *The New York Times* and revised an arti-cle denigrating Friedman's tuberculosis cure, which appeared in its Sunday edition just in time to greet him as his voyage from Germany ended in New York.

This saved Saranac Lake's tuberculosis cure industry and eliminated Friedman's chance to earn $1 million by curing 95 out of 100 tuberculosis patients.

Many physicians in Germany still use his preparation, which can still be obtained from SANUM-Kehlbeck Gmbh in Hoya, Germany. All of SANUM's medications are approved by the German pharmaceutical regulatory agency.

CANCER

A number of researchers have found that the origin of cancer is associated with pathogenic forms of fungi. They proliferate when the internal environment of our bodies deteriorates due to poor nutrition and exposure to carcinogens like cigarette smoke.

In his 1947 book *Siphonospora Polymorphia von Brehmer*, Dr. Wilhelm von Brehmer identified a blood parasite, a bacterial form of Mucor racemosus, a fungus, as a causal agent in the malignant growth of cancer cells.

Brehmer believed that carcinogens themselves don't cause cancer. Rather, they produce the environment that makes the dis-ease possible.

Enderlein developed cancer treatments using forms of these fungi, which neutralized the cancer causing bacterial forms.

In a double blind trial of a new cancer drug in Germany, a number of years ago, Utilin-S, a SANUM biological product, was given to half the patients. It was reported to be more effective than the new cancer drug.

Using a known treatment, instead of a placebo, against a new drug made sense to the Germans. Maybe we should see if it works for us in America. It seems to me that it would be more ethical to give human guinea pigs something that worked, instead of a "sugar pill," in drug trials.

When President Ronald Reagan underwent surgery for cancer, he was also treated by a famous German cancer expert, Dr. Hans Nieper, with biological remedies.

It will not be easy to introduce the use of these biological preparations into our medical care system, however. The barriers constructed by financial and professional interests are high and well constructed. It will be difficult for medical pundits to change what has become a well-established system, fed by pharmaceutical companies, financed by insurance companies and patrolled by malpractice attorneys.

It may require the intervention of Congress, which is still struggling to reduce the medical establishment's resistance to less revolutionary alternative therapies.

MULTIPLE SCLEROSIS

Multiple Sclerosis (MS) is characterized by the destruction of the myelin sheaths that insulate components of the nervous system and the formation of plaque in the brain and spinal cord. This disrupts the electrical control system of the body and ultimately results in permanent loss of nervous functions. The disease usually attacks young adults and leads to progressive disability. Common symptoms are loss of vision in one eye, double vision, difficulty walking, trembling hands, numbness or tingling.

Research indicates that the immune systems of MS patients may create antibodies that attack myelin. Because it rarely occurs in the tropics, where the sun tends to be intense, a connection with vitamin D, the sunshine vitamin, is suspected. The chance that an immediate relative of an MS patient will acquire the disease is about 15 times greater than normal.

Conventional medicine has established neither cause nor cure. Only palliative treatments are available.

According to Dr. Erik Enby, who wrote the book Hidden Killers, Enderlein discovered that MS is an infection that can sometimes be healed. He found that a number of different microbes present in the blood of MS patients were programmed to attack tissues of the nervous system. Enderlein developed medications capable of inactivating some of them, but did not live long enough to deal with all of them. Consequently the disease can be halted for some MS patients.

Enby reports success in halting the disease, for some MS patients, in his private practice with Quentakehl extracted from the mold fungus Penicillium frequentans, Nigersan made from Aspergillis niger, and Recarcin. These medications are said to destroy the pathogenic microbes and halt the disease process.

However, nerve damage already existing is not reversed so it is important to start treatment as early as possible.

Dr. Robert J. Rowen in the June 2005 issue of his *Second Opinion* letter reported that Dr. Linda Mattmann, Ph.D., discovered bacteria without cell walls in inflamed tissues. These are common in MS.

Rowen said he had success treating relapsing MS with the antibiotic Minocycline years ago in Alaska and was called a quack for using it. Now its use is accepted by the medical establishment.

These facts support Enderlein's finding that MS has a bacterial cause.

NOSOCOMIAL INFECTIONS

Harmful bacteria, which inhabit hospitals, become resistant to many antibiotics. When antibiotics are used to treat patients, they sometimes suppress susceptible bacteria and give the dangerous hospital bacteria a chance to thrive. These infections are said to affect more than 6 percent of hospital patients.

They are also responsible for costs that hospitals can't fully recover. It has been estimated that they add about $1 million annually for a 250-bed hospital.

Some European physicians, including Enby, prepare patients for their hospitalizations with injections of biological preparations like Mucor racemosis, using a SANUM product named Mucokehl and Sanovis, to create an internal environment capable of resisting pathogenic bacteria.

Enby claimed that patients with infections acquired in hospitals could be successfully treated with "Pefrakehl," made up of Candida parapsilosis, and "Notakehl," extracted from the green mold, Penicillum notatum.

ACQUIRED IMMUNE DEFICIENCY SYNDROME

The syndrome AIDS, which started out as an infection associated with homosexual males and intravenous drug users, is now widespread—a worldwide scourge. Robert Gallo of the National Cancer Institute, Lac Montagnier of the Pasteur Institute and Dr. Jay Levy of the University of California have all reported isolation of a virus responsible for this disease.

According to theory the human immunodeficiency virus, two thin strands of ribonucleic acid (RNA) wrapped in a protein coat, destroys T4 cells, important elements of the immune system. Without an effective immune system the AIDS patient is vulnerable to all sorts of diseases and ultimately dies because of this vulnerability.

Dr. Peter Duesberg, a retrovirologist and member of the Department of Biology of the University of California at Berkeley, named California scientist of the year in 1971, noted that the S in AIDS stands for "syndrome," which means that it is not a single disease entity.

In the March 1989 issue of the magazine *Cancer Research*, he also noted that even in dying AIDS patients it is very difficult to detect this virus and antigens to it. He also noted that many members of the AIDS research establishment have a financial interest in testing kits, and the companies that make and sell them, which can affect their objectivity.

Swiss biologist Dr. Bruno Haefli believes that growth forms of fungi may be involved. Juliane Sacher, M.D., working in the

1980s, stated that 85 percent of her AIDS patients remained stable or improved after 18 months of treatment with the wholistic treatments, Pefrakehl, Notakehl and ozone.

Enby also reported success in treating AIDS patients with Utilin and Nigersan. He found that patients improved dramatically and their blood picture under the microscope became healthier. He believes that the AIDS-causing microbe is pleomorphic and causes different diseases as it develops through different stages.

Dr. Karl Andreas Guischard used SANUM biological medications and ozone therapy and found that he could stop the progress of the syndrome. His attempts to interest University of Hamburg professors to initiate clinical studies were rebuffed, however. When he presented his results on 25 patients showing how blood factors normalized after 50 days, he was asked if he was sure the syndrome was AIDS. His results were declared unbelievable.

CHAPTER 5

POSSIBLE EPIDEMICS

Air transportation has opened the way for rapid dissemination of the disease vectors and a number of them have taken advantage of their opportunity. As a result, the West Nile virus and the Dengue viruses have made their way into new territories. New viruses generated in China, where birds, pigs, and people live in close proximity, require continual surveillance. Pigs accommodate both their own and human flu viruses and, as a result, provide for the transformation of bird flu into a threat to humans.

Lax enforcement of immigration laws has permitted the resurgence of old diseases in new and more difficult to treat forms. Legal immigrants are no longer required to pass physical exams before entry and illegal ones appear to be exempt from all controls. As a result, public health in some parts of the United States has reached third world levels.

WEST NILE VIRUS

The West Nile Virus (WNV) is widespread in the United States and may be already out of control, reports *Doctors for Disaster Preparedness Newsletter*. First identified by Tracey McNamara, a vet at the Bronx Zoo, in 1999 it has spread across the country.

Since most state health departments only test patients hospitalized with swelling of the brain or encephalitis—about 1 percent of those with WNV—the total number of victims remains unknown. However, in 2002 there were 4,156 cases of WNV and 284 deaths identified. Up to that time, this was the largest outbreak in North America of arboviral meningitis, an illness characterized by fever, headache, stiff neck and sometimes brain swelling that comes with WNV. In 2005, 6,000 cases were identified and there were a reported 139 deaths.

When the virus enters your bloodstream it encounters your immune system and, if that is in good shape and effective, it will be destroyed before it can do much harm. However, if it does manage to survive, it can enter your blood-brain barrier where the continuing battle with your white blood cells causes inflammation. The first action takes place in the membrane that envelops the brain and the spinal column. This is called the meningitis. If the virus gets through the three layers of this membrane you get inflammation of the brain. This is called encephalitis.

Mosquitoes spread WNV. They have become a serious threat aided by environmentalists, who feel that killing mosquitoes disrupts the food chain and, after all, "these diseases only kill the old and sick."

WNV is believed to be of Middle Eastern origin. Its route to the United States is unknown, and there remains a great deal of speculation—some bordering on the ridiculous—about its source.

The U.S. government sent samples to Iraq in 1985. A shady Iraqi defector once quoted Saddam Hussein as saying "laboratories outside of Iraq . . . free of UN inspection . . . will develop strain 141 of WNV."

Some believe the laboratories the defector referred to are in Cuba. A Russian reported that Cuba has one of the most sophisticated bioengineering laboratories in the world and Fidel and Saddam were friends. Other defectors say that the Cubans have studied the use of migratory birds to spread disease in the United States. It might take birds about three hours to fly from Cuba to the United States. However, Castro has many enemies. So some or all of this may not be true.

The Culex mosquito, a major vector in the spread of WNV, bites both birds and humans. In laboratory tests, 100 percent of crows die when infected. More than 100 species of birds have been infected. The house sparrow experiences a high viremia, an elevated presence of a virus in the bloodstream, for several days and is considered important in amplifying the spread of the virus.

Dozens of horses have been put down because of WNV in Cochise County, Ariz. Alpacas are infected in Colorado. Richard Bowen, professor of biomedical science at Colorado State University believes that cats, dogs and pigs, although they show no symptoms, also have a high rate of infection.

There is no specific therapy for this central nervous system disease, which results from infection with a single-stranded RNA virus, and causes neurological as well as movement disorders. A polio-like syndrome with possibly irreversible paralysis has also been noted.

Public health departments appear to be sitting back and watching the disease spread. Public information campaigns promote long sleeves, mosquito repellants and empty birdbaths.

Precautions such as pesticide use are discouraged and heavily regulated. Malathion is toxic to birds, fish and humans, and synthetic pyrethroids are toxic to fish and bees and may be a risk factor for some human diseases. However, neither has a residual effect.

There is no doubt that mosquitoes are carriers of dangerous diseases. There is reason to suspect that they may now, together with birds, be used in a biological weapon system.

The only claim of a cure for this type disease was made in 1971 in a paper available on the Internet. This therapy involves large intravenous doses of vitamin C, but it has never been thoroughly investigated because it is not part of the "standard of care" and consequently, physicians who might try it risk delicensure.

The use of large intravenous doses of vitamin C as a potential therapy for this disease needs to be investigated. The "standard of care" should not be allowed to interfere with people's health.

It is interesting to note that killer T cells in your immune system contain vitamin C. If deprived of this for any reason, they lose effectiveness. Vitamin C is one of the few anti-virals. Consequently it is reasonable to believe that intravenous vitamin C would be effective against this and similar diseases.

Additionally the banning of DDT as an insecticide needs reevaluation.

CARSON & DDT

A reading of Rachel Carson's 1961 book Silent Spring indicates that man's efforts to control insects have not always been successful and are sometimes counterproductive. It also appears obvious that these efforts have taken a toll on birds. Carson points out that life forms are adaptable and, by survival of the fittest, can render chemical controls ineffective. Many antibiotics have lost their potency against diseases through overuse. However, Carson's statements about DDT may not have been completely accurate.

According to Dr. J. Gordon Edwards of San Jose University, EPA trial judge Edmund Sweeney, after a seven months trial, issued an opinion on April 26, 1972, that DDT is not a carcinogenic, mutagenic or teratogenic hazard to man. He also decided that, under existing EPA regulations, it did not have a deleterious effect on freshwater fish, estuarine organisms, wild birds or other wildlife. He also stated that there was a present need for the essential uses of DDT.

There is evidence that the EPA convened a panel of experts to advise the EPA administrator, William Ruckelshaus, whose background was in law and politics, what he should do about DDT. This panel is reported to have advised him that DDT was useful and should not be banned.

The EPA administrator still banned DDT. When the ban was appealed, Ruckelshaus denied the appeal personally. It has been reported that the Environmental Defense Fund, which had filed the suit against the Department of Agriculture and the EPA resulting in the seven-month trial, influenced the administrator. Ruckelshaus was allegedly a member of the fund at that time.

As a result of the EPA ban on DDT many countries ended their use of it for mosquito control and malaria rates worldwide headed upward. Presently there are said to be between 300 and 500 million cases and 1 million deaths yearly due to malaria. A large proportion of the deaths are of pregnant women and young children.

INFLUENZA VIRUSES

In 1918, near the end of World War I, the influenza virus killed 20 to 40 million people in a 10-month rampage around the world. A new strain of the same virus came perilously close to staging a repeat performance in 1997. Quick thinking by a few scientists who convinced health authorities in Hong Kong to slaughter over a million chickens limited its human victims to six. They probably saved at least 20 percent of the population of the world.

Deadly new strains of this virus first described by Hippocrates in about 412 B.C. appear a few times each century. Nobody knows when the next one will arrive.

The disease spreads from person to person via coughs and sneezes. With the assistance of airplanes it could, potentially, spread itself around the world in hours.

Healthy, wild aquatic birds are the ultimate source. They carry it with them and spread it far and wide before humans are infected. In fact the virus they carry doesn't do well in humans. However when it is transmitted to domestic fowl and pigs via water contaminated by the wild birds' feces, gene sharing can produce extreme virulence.

Swine have receptors for both avian and human flu viruses and they provide the environment in which new strains of the flu develop. In China, birds, pigs and people frequently live close together and, as a result, most flu pandemics originate in China.

There are four different genera: A, B, C and Thogoto-like viruses, according to Robert G. Webster and Elizabeth Jane Walker. Their article "Influenza" in the March-April 2003 issue of *American Scientist* explains the biochemical aspects of influenza in considerable detail. Type A is responsible for pandemics.

Each year scientists identify two type A strains and one type B strain considered most likely to cause epidemics during the coming year. These are incorporated into vaccines. People who get the shots may be protected from these strains. They are not, however, protected from new strains, which may appear after the vaccines are in production.

Bacterial pneumonia kills thousands of elderly people every year. During an influenza epidemic, pneumonia mortality skyrockets. Vaccines are available for this disease, also.

Modern flu vaccines contain purified virus proteins, which stimulate the human immune system to create antibodies, which attack viruses containing these proteins. Weakened live virus vaccines are being investigated. Clinical trials of a nasal spray containing viruses indicate that it may be safe for both children and adults.

Anti-viral drugs such as Amantadine and Rimantadine have been used against the flu, but all strains appear to acquire resistance to them. Zanamivir and Oseltamivir are new drugs, which, if given early enough, can prevent replication.

Less than a dozen companies worldwide manufacture flu vaccines. There are only two in the United States and they have had difficulty meeting the demands of light outbreaks in the last few years. They would probably be overwhelmed if a virulent strain were to suddenly appear. Fear of potentially catastrophic lawsuits has driven some companies out of this business.

Since lawyers are not immune to this virus perhaps they should consider the advisability of resolving the legal liability problems of the virus manufacturers so that they, as well as the rest of us, can have a reasonable chance of surviving the next pandemic.

Drug companies take about 18 months to produce adequate quantities of antiviral drugs. Consequently such drugs would have to be stockpiled in advance of a pandemic.

Is there a program to take care of this contingency? No. All of the peoples of the world have common enemies in the form of bacteria, prions, viruses and yeasts. Many forget this fact in the heat of international competition for trade, oil, minerals, influence and religious preferences. Especially when war is contemplated it is tempting to consider how a temporary alliance with a bacterium or a virus might turn the tables on an opponent. In this connection there is much current discussion of anthrax and smallpox as weapons of mass destruction.

At the same time some of our common enemies like influenza and tuberculosis are major threats to all peoples. Bacteria are becoming progressively more resistant to available antibiotics and there aren't many new ones on the horizon. New viruses like Ebola continue to surface.

Hopefully the powers that be, in their haste to destroy one another and the artifacts of civilizations, will not forget that microbes and viruses are continuing threats to all the people of the world.

It is noteworthy that the influenza virus killed 500,000 Americans during the 1918-1919 epidemic, including 40 percent of the young men in the armed forces who died in the war. The virus was almost as effective as the German military.

Hopefully we will not experience another "lost generation" in this country. Perhaps the powers that be will remember that disease-causing organisms are continuing threats to all mankind.

DENGUE FEVER & DENGUE HEMORRHAGIC FEVER

Besides malaria and West Nile virus, dengue fever is another mosquito-borne threat to the world's people. With as many as 50 million cases worldwide, 2.5 billion people are at risk.

Dengue fever used to be considered a benign disease of visitors to the tropics. The first recorded epidemic occurred in 1879, over 200 years ago. Impeded by slow transportation, sailing ships, there were 10- to 40-year intervals between major epidemics. But the situation changed when a global pandemic originated in Southeast Asia after World War II. Today, the Centers for Disease

Control considers it to be the most important viral disease.

Four different dengue virus serotypes, isolated from one another, infested the tropics for a long time.

However, the geographical distribution of these four viruses and their mosquito vectors has expanded. And more frequent exposure to multiple serotypes resulted in concomitant outbreaks of both dengue and a frequently fatal complication called dengue hemorrhagic fever (DHF).

The first DHF epidemic occurred in Southeastern Asia in 1950. By 1975 it was a leading cause of hospitalization and death for children in many countries in that region.

Those who survive infection from a single serotype of the dengue virus are immune for life to that type and protected, for a short time, against the other three.

When only one type was present in an area, one infection meant that there were no further worries about the disease. Now a subsequent infection, by a different type of virus, opens up the possibility of DHF. If not properly treated, death rates from DHF can reach 50 percent. In addition, genetic variants of the four serotypes are working their way around the world.

The expanding geographical distribution of aedes aegypti mosquitoes, and the rapid rise in urban populations, is causing increasing numbers of people to be exposed to the disease. At the present time mosquito control is the only method available for countering these diseases. There is no virus-specific treatment. There are no vaccines, and there are not likely to be any for five to 10 years.

There is hope that a vaccine capable of countering all four dengue serotypes can be generated by a process called DNA breeding. This can be likened to directed evolution. Willem Stemmer invented this and co-founded Maxygen, Inc. to develop the technology.

Candidate vaccines have been successfully tested on mice, and trials with primates are following.

Before 1970 when use of DDT for mosquito control was widespread, only nine countries had experienced dengue epidemics. Now the disease is endemic in more than 100 countries.

With no new mosquito control technology available, the Public Health Department emphasis on disease prevention and community reduction of larval breeding areas is unlikely to have much of an impact.

After successfully eradicating mosquitoes from the United States and most of Central and South America with DDT our government chose to ban the insecticide, which had been keeping diseases like yellow fever, malaria and dengue under control. The mosquitoes must be rejoicing, because, by 1997, they were more widely distributed throughout the world than before the eradication program began.

A reevaluation of the decision to ban DDT ought to be seriously considered.

While there is only a small risk of a dengue outbreak in Texas and the Southeastern United States, northern Mexico and South and Central America have experienced epidemics of this disease. There were, for example, 390,000 cases in Brazil in 2001 including 670 of DHF.

With 40 percent of the world's population at risk of acquiring a progressively more dangerous disease, refusing to act would be morally unacceptable.

HOW TO STAVE OFF COLDS AND VIRUSES

A new disease, caused by what many are calling a corona virus, similar to common cold viruses, is threatening mankind. Originating in the Far East, its influence is spreading around the world with the aid of the airlines. At present there are only a hundred or so scattered casualties.

What can we do to protect ourselves?

Until something better comes along, our best protection is a strong immune system.

Many years ago Dr. Linus Pauling and his friends were recommending vitamin C—ascorbic acid—as the first line of defense against colds.

Pauling wrote a book *Vitamin C and the Common Cold*. It disclosed that ascorbic acid was one of the few substances with antiviral capabilities. He was ridiculed and vilified. In addition, he was denied any recognition for his work with vitamin C by his peers at the National Academy of Science.

The vestiges of this hostility remain in the form of the ridiculously low RDA for vitamin C, 60 mg. per day. This may be enough to prevent healthy young men from getting scurvy, but it certainly is not enough to give those of us—the majority of whom are not healthy young men—any real protection against disease.

The Tufts University Guide to Total Nutrition, published in 1990, states that although many people consume more than the RDAs; the RDA Committee has seen no evidence of benefits gained from over consumption.

I suspect that the committee didn't look very hard.

Dr. Sheldon Morgan of the University of California edited a nutrition book in 1997, which suggested that people should consider taking anti-oxidant supplements including vitamins C and E. Morgan also noted that your diet cannot supply enough vitamin E.

The fact that vitamins cannot be patented may have made it impossible to obtain the very large sums of money necessary for "gold standard" tests, which alone are convincing to the Brahmans of nutritional science.

Ascorbic acid is said to be most protective close to bowel tolerance levels. These levels vary from person to person with stress, injury and disease conditions.

Pauling said he took 18 grams per day. He reported that this provided him with protection against colds.

I usually take about 16 grams per day and I get very few colds. I have taken 52 grams in 24 hours after surgery without experiencing any difficulty from vitamin C.

In the past experts on the use of vitamin C recommended that people take relatively large amounts of it frequently at the first sign of a cold. I recall that, in 1939, a graduate student at a leading technical school dissolved 10 grams of ascorbic acid in a liter of distilled water and took 100 milliliters of the solution every half hour in order to get rid of an incipient cold before going to a party.

Other immune system enhancers have recently been advertised.

MGN-3, a proprietary product made from mushrooms, is one of them. It is reported to substantially increase the activity of killer T-cells.

The influenza virus killed millions of people, including a substantial part of our Army, near the start of the last century. Let us hope that the new virus will be less effective.

PUBLIC HEALTH ENDANGERED BY IMMIGRANT DISEASES

A recent communication from Doctors for Disaster Preparedness' claims that Great Britain is again leading the world, this time by reintroducing diseases it once almost eliminated.

Along with the admission of about 150,000 legal and medically untested immigrants, it has imported significant amounts of tuberculosis and hepatitis B. Presently one London borough has a tuberculosis rate higher than that in China. Ninety-five percent of the Hepatitis B Cases are said to be imported.

While we in America have the Immigration Act of 1891, which permits us to refuse to admit aliens dangerous to the public health, our political establishment does not enforce it. As a result about 300,000 untested aliens, and possibly a much larger number, enter our country every year. About 8 million illegals are already imbedded, carefully avoiding contact with government officials.

In 1987, when immigration amnesty began, tuberculosis was expected to be eradicated in the near future. Soon after that the current upsurge in tuberculosis cases began. The 2.1 million illegals, which suddenly became legal, undoubtedly contributed to this upsurge.

There are tens of thousands of Chinese immigrants who enter illegally through Canada, which has a more porous border than Mexico. They probably come from the 250 million young men in China who cannot find wives there. Along with them they could be carrying an increased risk for an epidemic of severe acute respiratory syndrome, known as SARS.

Third world health conditions are said to already prevail in parts of the United States such as El Paso and San Diego where leprosy can be found. And along the southern border of Texas where up to 50 percent of the children have hepatitis.

Our immigration policy is a shambles. Both Democrats and Republicans have made their contributions in their ceaseless efforts to garner new voters for their causes and, to make governing easier, by creating new voting blocs to play one against the other.

Progress in public health in the United States, which has been going on steadily for over 100 years, is now endangered by the federal government's failure to enforce its own immigration laws.

CHAPTER 6
NUTRITION

Our involvement with the World Trade Organization may easily become the means for the pharmaceutical industry and the medical establishment to hamstring the American supplement industry and possibly alternative medicine. Europeans have already had their access to supplements sharply curtailed. In 2004, bills were already in place in both the House of Representatives and the Senate to implement the *Codex Alimentarius* which was developed under United Nations sponsorship over the last several years.

The purpose of the *Codex* is ostensibly to protect the world's people, but the limitations placed on the number of substances to be made available as well as the quantities of them are likely to have negative impacts on public health. For example, most supplements will only be available in RDA amounts which are considered to be inadequate for the maintenance of optimum health. Limiting vitamin C to this level will undoubtedly increase the need for medical services and prescription drugs. Too small amounts of folic acid will probably increase the rate of heart attacks.

Soils on our farms have been depleted of minerals which used to be present. Foods raised on these soils are less nourishing than they used to be. Additionally, foods processed for long shelf life are generally less nutritious and where hydrogenated oils are included they can be detrimental to health. High fructose corn syrup, which is ubiquitous in processed and packaged foods, has also come under suspicion as a potent health hazard, along with sugar.

The food pyramid, with its focus on carbohydrates, may be one of the reasons that there are so many obese people. It is possible that this focus was warranted in by-gone days when most people did hard manual labor on farms, but very few people work on farms today and most are not physically active. Watching TV, playing video games or surfing the internet has taken the place of outdoor exercise.

CODEX ALIMENTARIUS

A coordinated assault on another freedom, the freedom to purchase and use dietary supplements, is planned for 2005. In that year the *European Union Directive on Dietary Supplements*, part of the *Codex Alimentarius*, takes effect. This directive severely limits access to more than 100 percent of RDA amounts of 15 minerals and 13 vitamins and completely eliminates access to other minerals important for health.

This is not only a disaster for Europeans; it is also a disaster for Americans. Thanks to our fearless leaders, our World Trade Organization membership makes enforcement of the European rules mandatory on our government.

To make enforcement of the European rules consistent with American law, two bills have been introduced into the legislative process: S. 722, the Dietary Supplement Safety Act, and H.R. 3377, the Dietary Supplement Access and Awareness Act. If these laws are passed the powers of the FDA and the medical Establishment will be enormously increased, and Americans will lose the right to control their health.

Some Europeans are fighting back, however. The British Alliance for Natural Health's attorneys won its case to have the European directive overturned in the High Court of Justice in London. An appeal is being referred to the European Court of Justice.

Meanwhile, the major American newspapers have been generally silent about this threat to the country's well-being, even though it has been big news in Europe for some time.

Perhaps some of the mainstream media's advertisers, like the pharmaceutical companies, wish to see the directive go into effect because it would increase their business.

The medical establishment has been concerned about the increasing numbers of people who are turning to alternative therapies and dietary supplements. Undoubtedly, the directive and the proposed bills in Congress met with their hearty approval.

Dr. Jonathon Wright notes that a woman with dangerously elevated homocysteic levels will no longer be able to take five milligrams of folic acid, 100 milligrams of B6 and 500 milligrams of B5. Instead, she will only be allowed to take one milligram, 10 milligrams and 200 milligrams respectively, leaving her more likely to have a heart attack.

It should be noted that RDAs represent the amount of essential vitamins required to prevent diseases like scurvy and pellagra from affecting young men. But most of us are not young men, and many of us are not completely healthy.

Citizens concerned about their health should notify lawmakers that these bills and the European directives on dietary supplements are not acceptable and will warrant a vote for their opponents if not quashed before the election.

Also, Americans should inform congressmen and senators that the World Trade Organization interfering with our ability to take care of our own health and well-being will not be tolerated.

Rep. Ron Paul (R-Tex.) has proposed H.R. 1146, the American Sovereignty Restoration Act, to restore the Constitution to its rightful position as the supreme law of the land.

VITAMIN C AND CARDIOVASCULAR DISEASE

Almost two-thirds of the population of Western developed countries suffer from atherosclerotic deposits, or "blockages," in their arteries. But a combination of vitamin C and the amino acid, lysine, could help alleviate the problem inexpensively and effectively, according to important research by a noted scientist.

Atherosclerotic deposits consist of fatty materials, proteins and minerals, particularly calcium. They narrow the arteries through which blood flows. The narrowed arteries are said to be occluding with plaque. The calcium associated with plaque appears to anchor it in place.

For more than 30 years physicians associated with the

American College for the Advancement of Medicine have been removing the calcium with a substance called ethylenediamine tetracetic acid in a process called Chelation.

With the anchor removed the plaque tends to disappear. Of course it comes back if the patient doesn't change the lifestyle which caused it in the first place.

If a blood clot enters an occluded artery, it can block blood flow and kill some of the tissue associated with the artery. This is called a heart attack, if the artery is feeding the heart, and a stroke, if it is feeding the brain. Heart attacks and strokes are associated with high cholesterol.

Cholesterol is an essential ingredient in every cell of your body. Without it your cells would not be able to adapt to constantly changing conditions in your body fluids. In your skin cholesterol is converted by ultraviolet light from the sun into vitamin D which you need to maintain strong bones.

Cholesterol is the raw material from which male and female hormones are made. During pregnancy the placenta converts cholesterol to progesterone which keeps the pregnancy from being terminated.

The liver turns cholesterol into bile, which you need to digest fats and oils. Obviously cholesterol is a very important material and you need adequate amounts of it to stay alive and well.

Cholesterol cannot be metabolized to carbon dioxide and water. It is either removed with bile in the stool or recycled. High fiber diets help to remove it. In the absence of dietary fiber over 90 percent may be recycled.

Most people think that there are several different kinds of cholesterol. Actually it is a well-defined chemical substance with a complicated structure. It moves through your blood stream in various lipoproteins.

Lipoproteins are globules of fatty material surrounded by a protein shell mainly composed of a substance named apoprotein B-100. Very low-density lipoproteins (VLDLs) are made in your liver and are loaded with triglycerides, which your cells use for fuel. They contain about 20 percent cholesterol.

Low-density lipoproteins (LDLs) carry fats, fat-soluble vitamins and cholesterol to your cells. They contain over 40 percent cholesterol. High-density lipoproteins (HDLs) carry fats and cholesterol back to your liver for conversion to bile. They contain over 20 percent cholesterol.

A new member of the lipoprotein family was discovered in 1962. It is called lipoprotein (a) Lp(a). It consists of LDLs with a substance called apoprotein (A), apo(a) attached to apoprotein B-100. It is this substance that thickens arteries in conjunction with fibrin and fibrinogen. The concentration of Lp(a) in your blood is usually low when the concentration of vitamin C is high, and vice versa. It plays an important role in Dr. Linus Pauling's theory of cardiovascular disease (CVD).

Most animals make their own vitamin C from glucose. They don't get scurvy and they don't have Lp(a) in their blood.

According to Dr. Pauling, we would be making between 1 and 20 grams per day of vitamin C if our ancestors hadn't lost this capability about 40 million years ago.

The equivalent of 3.18 grams per day for a 175-pound man prevents the development of atherosclerotic lesions in the guinea pig.

Dr. Pauling believed that a condition, which he called pre-scurvy, develops when people obtain less than adequate amounts of vitamin C in their diets.

Scurvy is a disease which results from extreme depletion of vitamin C. Capillaries become increasingly fragile in this disease, and this leads to massive hemorrhaging and death.

In pre-scurvy, less than optimum amounts of vitamin C in the diet causes weakened arterial walls which the body's systems attempt to strengthen with the aforementioned Lp(a). The attached apo(a) is an adhesive material which normally mediates organ differentiation and growth but, when vitamin C levels are too low, it increases the stability of the blood supply system by patching weak areas, in conjunction with foam cells calcium and other substances. Unfortunately these patches, plaques, impede blood flow causing CVD.

According to Dr. Pauling, all you have to do to reverse this condition is to increase your intake of vitamin C and lysine an essential amino acid. The lysine counteracts the effects of the substances responsible for weakening your arteries, mainly plasmin, and the vitamin C provides raw material for rebuilding the collagen based arterial walls.

He authored three papers which described successful use of lysine and vitamin C to relieve the effects of CVD.

The first described the experience of a biochemist who had been told by his cardiologist that he was not a candidate for a fourth by-pass surgery because he no longer had useable veins.

After about six months, during which he took gradually increasing amounts of lysine—his maximum was six grams per day—he was able to work and take long walks without his usual severe anginal pain.

In 1991 Matthias Rath, M.D., and Dr. Pauling submitted a paper for publication in the *Proceedings of the National Academy of Science*, titled "Solution of the Puzzle of Human Cardiovascular Disease: Its primary cause is ascorbate deficiency, leading to the deposition of liporprotein (a) and fibrinogen/fibrin in the vascular wall."

It was accepted for publication and then rejected. Subsequently the officials managing the academy went out of their way to insure that Dr. Pauling would get no recognition from them for his work on vitamin C. Albeit, he was honored for his other contributions to science.

If Rath and Pauling were correct in their analysis of CVD, as described in the above paper subsequently published in the *Journal of Orthomolecular Medicine* in 1991 and elaborated on in "A Unified Theory of Human Cardiovascular Disease Leading the

Way to the Abolition of this Disease as a Cause of Human Mortality" published in the *Journal of Applied Nutrition* in 1992, then it would seem that a great many of us may have died before our time in the last 10 years.

If you wish to try out Dr. Pauling's theory, please consult a health professional before you start. CVD problems are generally serious and may be life threatening.

HOW MUCH VITAMIN C DO YOU REALLY NEED?

Many people do not realize that man is one of the few mammals which does not manufacture its own supply of vitamin C, also known as ascorbic acid.

The rates of production by a number of different animals like the cat, dog and goat have been measured. The heavier the animal the more vitamin C it produces.

However, a 154-pound man would need to produce between 1.75 and 3.50 grams per day to keep up with the other animals.

There are experts who feel that taking more than 140 milligrams per day of this vitamin is wasteful because, for many people, more than this amount results in urine containing the vitamin—"Expensive Urine."

Loading tests have shown that 20 to 25 percent of a 1 gram per day dose shows up in the urine within 6 hours. When much larger doses are taken, as much as 62 percent can show up within hours.

Generally speaking, the recommended daily allowances have been based on the amounts needed to prevent scurvy in healthy young men.

But healthy young men represent only a small part of the total population.

To my knowledge, no one has determined how much of any vitamin is required by unhealthy old men—or women.

The results of a test involving 88 patients, half being schizophrenic, are reported in Dr. Linus Pauling's book *How to live Longer and Feel Better.*

Each was given 1.75 grams of vitamin C by mouth. During the following 6 hours each patient's urine was collected and then analyzed.

The amounts excreted varied from 2 percent to 40 percent of the amount ingested. The mental patients excreted about 60 percent less than the others.

This clearly indicates that the need for vitamin C is quite variable. Just as one size shoe doesn't fit every foot, the recommended daily allowance is not going to fill everybody's need for this essential vitamin.

Incidentally, excreted vitamin C is not a complete loss. According to Dr. Pauling, it protects against urinary tract infections.

Large doses can have a laxative effect causing looseness of the bowel. This is said to be greater when the vitamin is taken on an empty stomach.

Dr. Pauling recommended using this laxative effect to reduce chances of developing colon cancer. Additionally, the vitamin appears to be most effective in fighting disease when the dose is close to the bowel tolerance limit.

Vitamin C is one of the few naturally anti-viral materials. There has been much controversy about its usefulness against colds and influenza.

It is notable that Dr. Pauling's interest in vitamin C was sparked by his observation that he and his wife experienced a striking decrease in the number and severity of colds after they started taking large quantities of it. I have enjoyed the same benefit.

The effects of the vitamin are dose related. Many tests at low dosages during the early years of this controversy showed little effect except reductions in the duration of symptoms.

However, physicians experienced with this therapy recommend intakes near the bowel tolerance limit, said to be between 4 and 15 grams per day for people in good health.

Dr. Irwin Stone, a pioneer in the use of vitamin C, recommended taking 1.5 to 2 grams by mouth at the first sign of a cold and repeating the dose at 20 to 30 minute intervals until symptoms disappeared. He said this usually happened by the third dose.

Collagen is the material that holds your body together. Vitamin C is destroyed in the process of producing collagen. If your skin is cut, collagen is generated to make scar tissue.

It seems obvious that the more extensive the damage, the greater the amount of vitamin C needed for the repair.

There are many references in the scientific literature attesting to the efficacy of vitamin C in wound healing.

When I had a need for a hernia repair, I put the theory to a test. I increased my intake of vitamin C prior to the surgery and took 52 grams of it in the 24 hours after it at the rate of about 2 grams per hour.

I experienced no digestive difficulties, my wound healed well and I did not need the prescribed pain medication.

I was careful to reduce this high intake over the next 3 days to avoid a rebound effect. If you take a lot of vitamin C and stop taking it suddenly, your liver will take it out of your immune system leaving you vulnerable to infection.

It is interesting to note that your bones are made up of layers of collagen and the mineral apatite which together form a matrix of semiconductors. It is reasonable to believe, therefore, that vitamin C is also of value in maintaining the integrity of your skeleton.

POISON?

There are those who say that large quantities of vitamin C are poisonous.

Dr. Pauling reported that he took 18 grams of it every day and he lived to be over 90.

In bulk, vitamin C is not expensive. It costs about 3.5 cents per gram. Thus 1.75 grams costs about 6 cents and 18 grams about 65 cents.

Of course your body needs other supplements to stay healthy.

Dr. Joel Wallach has stated that there are 60 minerals, 16 vitamins, 12 essential amino acids and 3 essential fatty acids that your body requires in order to prevent dietary deficiency diseases.

The soils in the United States have been deficient in minerals for years so you cannot get everything you need from food alone.

In 1994, Dr. Wallach stated that the life span for the average American was 75.5 years, but was only 58 years for medical doctors.

An attempt to update the life span for physicians was unsuccessful because a representative of the American Medical Association stated that they no longer keep these statistics.

Possibly the physicians who have been advising their patients to avoid "Expensive Urine" have been taking their own advice—and dying early of "Cheap Urine."

ASTHMA EMERGING AS DEADLY THREAT TO KIDS

Kids from ages four to 15 have been dying from asthma at progressively increasing rates since the late 1970s in the United States and no one in the medical establishment can explain why. A new look at sensitivities to certain types of food may provide insight into this life-threatening disorder that is affecting more and more children every year.

In 1994, the rate for asthma was 3 times higher than in 1975, according to a medical expert from the Jewish Hospital in Denver, Colo., who could not give a reason for the rapid increase. In 1999 it was 4 times higher.

Loyola University Health Systems reports that almost 5 million children under the age of 18 had asthma in 1994 and there were 159,000 hospitalizations of youngsters under 15 for asthma in 1993.

In the 1950s, asthmas-related deaths were relatively high. However, the number of fatalities decreased to a minimum in the late 1970s. But, since then, deaths have been increasing steadily.

Today, it is said to be the number one respiratory disease in America. Some 50 percent of the cases are mild, but 10 percent are severe and life threatening.

Boys are more likely to become asthmatic than girls and African-Americans are more likely to die of it than Caucasians.

What is causing this epidemic?

There is no simple explanation, but many blame changes in our lifestyle, the food supply and indoor and outdoor pollution.

In 1990 about 10 percent of our food supply was refined and chemically enhanced. Now about 90 percent is. Over 3,000 different additives are being added to our food supply and over 10,000 contaminants are said to be derived from pollutants. Additionally some essential oils are removed to improve shelf life.

Obesity, which makes people prone to asthma, is believed to be encouraged by lack of exercise and the standard American diet—SAD—full of simple carbohydrates, which are high-glycemic food.

Adults with a Body Mass Index over 28 are 2.4 times more likely to develop asthma and are at a higher risk for emphysema as well. In recent years more and more people are beginning to resemble the food pyramid—much larger at the bottom than the top. Even small children are taking on these proportions.

Dr. Fred Pescator, author of the book *Asthma and Allergy Cure*, claims that he was a fat kid with asthma and allergies. After he dieted to lose weight, cut down on high glycemic foods, and concentrated on hamburgers and salads with olive oil dressing he claims that his asthma and allergies disappeared.

Excessive use of antibiotics kills not only harmful bacteria but also the good kind, which we need to digest our food. If the good ones are not replaced quickly enough, candida, a yeast that plays a role in digestion, can get out of hand. It feeds on sugar, which the average American consumes at a rate of about 138 pounds per year.

Pescator claims that people with allergies and asthma improve dramatically once yeast is removed from their bodies. However, this syndrome is not taught in medical schools.

Seventy-eight percent of asthma patients have allergies, which can exacerbate asthma. Allergies are the most common of the chronic diseases of Americans; they are more prevalent than heart disease or diabetes.

According to the Food Allergy Network and *The Source Book on Asthma* 70 percent of Americans believe they have food allergies—but they are wrong. Only 1 percent to 2 percent of people are actually allergic to food. According to the Asthma and Food Allergy Foundation, what the rest of these people have may be food sensitivities, but this is a controversial area in the research phase.

Pescator believes that there really are food sensitivities. People who have them, he says, are being ignored by traditional physicians who believe that they don't exist. "Food sensitivities may be the most under diagnosed medical problem in American history," said Pescator.

Dr. James Braly's book, *Food Allergy and Nutrition Revolution*, provides a different viewpoint.

Braly claims that food sensitivities are really food allergies which are mediated by a different element of the immune system, immunoglobulin G (IgG) instead of immunoglobulin E (IgE) which conventional physicians associate with allergies where there is an immediate reaction to allergens.

Typically, IgG reactions to allergens are delayed. They may appear a few hours after a food is eaten or perhaps several days afterwards. People can live for years with this kind of allergy and never know it.

Food sensitivities, or IgG mediated food allergies, can have widespread effects not only on asthmatics but also on people with many other diseases such as multiple sclerosis, arthritis and gastrointestinal reflux disease (GRD). It is interesting to note that 45 to 65 percent of asthmatics have GRD. It makes asthma more difficult to manage.

Special immune system tests are required to determine which foods are causing sensitivities or IgG-based allergies.

Braly relies on the ELISA test, a blood test that checks your immune system reaction to about 100 different foods. I had one and found that I was allergic to 18 different foods. I ran a test and found that I could recognize delayed symptoms from eating some of them.

Pescator has a different method of detecting sensitivities via immune system based tests.

CHILDHOOD DEFICIENCIES

Childhood food allergies became more prevalent during the past century as breast-feeding became less common. Experience indicates that four to six months of breastfeeding can substantially delay the appearance of childhood food allergic diseases such as middle ear infections, asthma, chronic diarrhea and insulin-dependent diabetes according to Braly.

Early exposure of infants to foods other than mother's milk has resulted in a high rate of rice allergies in Japanese children and cow's milk allergies in bottle-fed Scandinavian and British children. The delicate balance between friendly and unfriendly bacteria can be upset in this way and impaired digestion of food can result. Inappropriate and overuse of antibiotics can also cause health problems which pro-biotic supplementation can help to correct.

Deficiencies in essential fatty acids (EFAs), which are important components of the membranes surrounding all cells in the body play an important role in the development of allergies. Not enough in the diet can cause weakening of the cell membranes and increase permeability. This helps allergens in the environment to penetrate the mucosal linings of the airways, the skin, and the gastrointestinal tract. Studies show that hyperactive children from families with a history of hay fever, hives, asthma, and eczema are usually either deficient in EFAs or unable to metabolize them properly.

Food allergens produce inflammation in various parts of the body when incompletely digested allergic food particles pass through the walls of the digestive system and get into the blood stream. If not cleared by the immune system, they are deposited in various tissues. Then chemical mediators like histamine, prostaglandins, bradykinins, and leucotrienes cause inflammation. When deposited in the joints, these particles can initiate and exacerbate arthritis.

Mucosal permeability and immune system malfunctions have been blamed on free radicals. In fact, rheumatoid arthritis sufferers have experienced marked improvement when vitamins A, B-5, C and E plus selenium are included in the treatment program. Food allergy elimination and anti-oxidant and EFA supplementation make for a powerful anti-arthritis therapy, according to Braly.

Vitamin C is the most abundant anti-oxidant in the inner lining of the lung. It is believed that, when more is consumed, more is available to inactivate contaminants as they enter the lungs.

Drs. Trenge and Koenig at the University of Washington reported that 500 mg. per day of vitamin C with 40 I.U. of vitamin E benefited patients who were sensitive to air pollution. Braly suggests that children with asthma be given 500 mg. of vitamin C before exercising. Recent tests indicate that 2,000 mg. is even more effective.

Allergic reactions to the following foods are common: corn, yeast, wheat, rye, milk, cheese, eggs, soy beans, coffee, oranges, chocolate, tomatoes, white potatoes, spices, malt, peanuts, beef and pork.

Signs of food allergies in children include craving or demands for a particular food, dark circles or swelling under the eyes, eye wrinkles (Dennie's sign), horizontal crease across the nose, migraine headaches, excessive coughing after exercise, nightmares, paleness without anemia, bad breath, ear infections, tinnitus and hyperactivity.

It is not uncommon for people to become addicted to allergic foods. Attempting to cope with allergens, our bodies release a variety of hormones and some compounds, which can produce a temporary "high" often accompanied by relief from discomfort. But depletion "lows" follow. After a number of cycles people begin to feel miserable without the food. Physiological addiction, craving for the allergic food because of the temporary relief gained by eating it, follows.

It is said that some addictions to alcohol are caused by allergens from the materials used in the manufacture of the beverages such as malt, yeast, corn, etc.

If your child has severe food allergies, according to Pescator, you should:

• Be prepared for emergencies—have a plan.

• Be very careful shopping—don't buy foods containing the allergen.

• Carry an Epi-pen—an injector with disposable cartridge containing epinephrine.

Exercise can trigger an asthma attack. If asthmatic children feel warm or are flushed before exercising, or unusually fatigued or light-headed, he or she shouldn't exercise that day.

Other reasons for higher incidences of asthma can include the fact that with the advent of television, video games and the Internet, today's children are spending a lot of time indoors in sedentary activities. Concurrently, increasing costs of energy have encouraged parents to minimize heating and cooling costs. In many cases, outside air flow into homes is minimized as a result. In some instances indoor pollution is said to be worse than outdoor pollution. Increasing asthma death rates may also be a consequence of too many immunizations.

According to Dr. Buttram of Woodlands Research Center, Quakertown, Penn., the 22 or more immunizations, which are effected by direct injection into the bloodstream, eliminate challenges to the mucosal immune systems. This leaves them undeveloped and less able to cope with potential allergens in later years.

Four studies found that vaccinated children had more asthma and atopic diseases than controls. Authors of one of these studies noted that the 30-year-long increase in allergic diseases among UK children remains unexplained.

They hypothesized that infections in early childhood may prevent allergic sensitization and that this protection has been lost as exposure to disease early on has diminished.

Asthmatics shouldn't eat known allergic foods before exercising. Nor should they exercise until at least 3 hours after eating. They should take 500 to 2,000 mg. of crystalline vitamin C an hour before exercising and warm up before and cool down after the exercise. Braly suggests 30 to 40 minute workouts, although exercising for shorter periods interspersed with rest is also acceptable.

The following are recommended supplements for asthmatic children, according to Braly: multivitamin capsules 1 per meal; multimineral capsules 1 per meal; vitamin C, 1,000 mg., 2 times per day with meals; evening primrose oil 3 capsules 3 times per day with meals; max EPA 2 capsules 2 times per day with meals; vitamin A micellized 1 drop every other day; vitamin B complex 1 with lunch; vitamin E micellized 10 drops per day; and pantothenic acid, 100 mg, 1 with dinner.

Our children are our future. Threats to our future need prompt attention.

THE ATKINS DIET

After many years of controversy, the weight loss diet promoted by the late Dr. Robert C. Atkins has finally been tested, and the results, published in *The New England Journal of Medicine*, proved that his predictions were correct. Unfortunately, he is not around to savor the fruits of victory since Atkins died on April 17th of this year.

Beneficiaries of his low-carbohydrate diet lost weight; their blood triglycerides fell, insulin sensitivity improved, high-density lipoproteins (good cholesterol) increased and total cholesterol either stayed the same or decreased.

There is a simple principle underlying the Atkins diet: to lose weight your body must burn stored fat. Most Americans eat a lot of carbohydrates and drink a lot of them, too, in the form of soft drinks loaded with sugar. Their bodies preferentially use some of them for energy and store the rest as fat. The consequences of this are apparent all around us. We have an outbreak of obesity.

Primatologists in the field have a technique for determining the adequacy of the food supply of the animals they study. They spread a plastic sheet over a spot where the subject will relieve itself. Then they test the collected urine for ketones. If ketones are present, they know that the animal was using some of its own stored fat.

Atkins recommended that dieters use the same technique, looking for ketones in their urine, to ensure that they were burning their stored fat.

In essence the Atkins diet requires the dieter to cut back on carbohydrates until ketones show up in the urine to ensure that the fat-burning mechanism is turned on. Subsequently, carbohydrates can be added, but the key to weight loss is using fat for energy.

Jane Brody, in a recent *New York Times* article, questioned the safety of diets high in saturated fats. But Eskimos used to survive in the arctic by eating mainly protein and fat because carbohydrates were not readily available.

In the Eisenhower administration, his personal physician, Dr. Paul Dudley White, recommended that all Americans cut back on their consumption of meat and fat after the president's heart attack. He later retracted this recommendation based on a year-long investigation of two arctic explorers who questioned his advice. They ate nothing but meat and fat, half the year in a hospital under close observation, and demonstrated no adverse effects. White wrote an introduction to the book in which they related their experiences during the yearlong test.

Many carbohydrates, like sugar, can be addictive for some people. These substances can cause overeating with consequent deposition of fat. Fats, on the other hand, tend to discourage overeating by producing a feeling of fullness.

Some people also have food sensitivities or delayed action allergies, which can cause inflammations, which tend to draw and hold water in their vicinity. Eliminating such foods results in weight loss when this water, no longer needed, passes out of the body.

The hard part of cutting back on carbohydrates is recognizing and eliminating your addictions. Total withdrawal for a short period of time has worked for some people. Recently, a physician friend disclosed that he had lost 20 pounds on the Atkins diet and expected to lose more.

CHAPTER 7
MISCELLANEOUS TOPICS

From time to time I felt impelled to write short articles about a variety of subjects and they are all included here. Their main focus was on health care, but several dealt with the impact of environmentalist efforts to promote the interests of predators. In some parts of the country, man is no longer at the top of the food chain. It is unfortunate that those who promote these ideas are not generally exposed to the risks and hazards that farm animals and children who live in rural districts face.

The lone superpower status of the United States appears to be threatened by the emergence of new combinations of countries. The balance of power principal was actively promoted by Great Britain during its long history in order to prevent the emergence of a single dominant country in Europe. Russia is taking a leaf from England's book.

Brain research has been around for a long time, but recent efforts to explore capabilities and recover waning abilities are promising.

Crop Circle phenomena have been the subject of much discussion and serious efforts have been made to discredit research in this mysterious area. However, I believe that the witness whose experience I have recorded is credible. It is quite likely that modern science still has much to learn and a great deal of it probably involves energies of which it presently has no knowledge. After all, the lowly chicken is able to violate the second law of thermodynamics with impunity.

Last, but not least, God's existence is the final, appropriate subject which ends this book.

ANTI-MALIGNAN ANTIBODY SCREEN TEST

The new Oncolab, Inc. Anti-Malignan Antibody Screen (AMAS) blood test can detect all kinds of cancer, with 95 percent accuracy and a false-positive rate less than 1 percent—long before the disease is detectable using conventional techniques.

During malignant transformation of cells in the human body an antigen, malign in nature, is formed and, in response, the immune system generates the anti-malignan antibody.

There is a normal concentration of the antibody in human blood, which increases with age, paralleling the risk of developing cancer. Early in the development of all human cancers, this concentration increases markedly above normal.

The antibody is extremely toxic to cancer cells. The greater the concentration in the blood of cancer patients, the better their chances for survival.

The structure and function of both antigen and antibody have been determined. This provides a sound theoretical basis, not only for a useful screening test, but also for the possible development of vaccines as well as methods of locating malignancies. It may also lead to methods of stimulating the human immune system to take care of cancerous growths without the use of toxic medications.

Because of the extreme sensitivity of the antibody to the initiation of malignancy, it promises to be particularly useful in early detection of all cancers. Its immediate usefulness for breast cancer screening can be easily demonstrated.

Mammography has been used extensively in this country for the detection of breast cancer. The National Cancer Institute says that, in 1997, 65 percent of American women over 40 had a mammogram within the previous two years.

They claim that this screening reduces the risk of dying from breast cancer by 30 percent for 50 to 69 year old women. Many women feel that it has saved their lives.

However, recent reviews of available data by Dr. Peter C. Gotzsche and Dr. Kirsty Loudon Olson, of the Nordic Cochrane Center located in Denmark and Norway, led to the conclusion that the procedure has not reduced mastectomies significantly, nor has it had much of an effect on breast cancer death rates.

According to Dr. W.C. Douglass, mammograms can only detect tumors after they are large enough to be advanced cancers. Thus a dangerous cancer can develop unchecked because it is too small to be detected—a false negative.

Not all large tumors are cancerous, and the risk of false positives is over 40 percent after nine mammograms, and rises to 49 percent after 10. False positives cause unnecessary trauma and medical expense, and frequently result in compromised immune systems resulting from the removal of lymph nodes.

While the AMAS blood test is not sufficient for diagnosis, it can help physicians identify cancerous conditions earlier and more accurately than is possible with mammography alone.

If cancer is present, it will still have to be located. If found to be in the breast, mammography would be useful.

While 37,000 women and 400 men are expected to die of breast cancer in 2002, the lifetime risk of the average woman getting the disease is only about 11 percent.

A daily dose of 200 micrograms of Selenium can reduce the risk of getting cancer.

In the January 2002 issue of *Real Health*, Dr. Douglass states that there is no need for quick and drastic treatment if the AMAS test detects early cancer cells because most cancerous tumors are slow growing.

GRIZZLY BEARS SUPPLANT MAN AT TOP OF FOOD CHAIN

On June 7, a rancher in northern Montana went on horseback to see if he could recover some of his cattle which had been caught in the 10-foot drifts of a heavy snowstorm. Unfortunately his animals had not survived. They were among the approximately 2,000 carcasses spread around the nearby fields.

But it wasn't necessarily the snow that killed the rancher's herd. Much to his surprise, about 40 grizzly bears were enjoying what they considered to be a windfall in an area where only a few of them were supposed to be present.

The feeding grizzlies did not wish to be disturbed and one drove the rancher off. These bears are able to outrun an elk or a man on horseback. This one was catching up to the man on horseback but, by fortunate chance, his camera, which he had been carrying to report dead cattle, fell to the ground, distracting the bear and allowing him to escape.

Grizzly bears are not the only large predators that the U.S. government has reintroduced into Montana. Wolves have been planted in Yellowstone and Glacier National Parks.

Fifteen years ago there were 20 wolves in Montana. Now 200 have been reported to be living there. They are wide-ranging animals and now there are three packs in the Helena area, which is about 175 miles from Yellowstone.

The Independent Record of Helena reported a growing number of ranchers are finding dead and maimed livestock in areas where wolf packs are operating.

One family watched through the windows as wolves ran through their yard. They reported finding one of their sheep eviscerated with its hindquarters partially eaten by wolves. Recently they found three of their calves dead. Still, they do not have enough evidence to prove that wolves had killed them.

The Fish Wildlife, a Montana organization, may reimburse ranchers if they can prove that wolves killed their stock—for example $125 for a sheep. But there are no provisions to pay for breeding stock. Some bulls bring a price of over $50,000 at auction.

The U.S. Fish and Wildlife Service (FWS) and the Department of Agriculture Wildlife Service are supposed to take care of problem predators. They sometimes catch, collar and relocate "bad" wolves. In extreme cases they shoot them.

Between 1987 and 2002, 44 wolves were killed and 37 moved, according to the *2001 Gray Wolf Recovery Report* issued by FWS. During the same period 91 cattle, 68 sheep, 10 dogs and four llamas were confirmed wolf kills.

Periodic aircraft surveys covering 200 square miles of the 98 million acres of Montana indicate that the elk population is stable, deer are at a periodic low, but grizzly bears are increasing slowly and wolves are increasing rapidly.

Recently the FWS proposed to reclassify wolves from endangered to threatened. With this classification, ranchers would be able to shoot wolves seen attacking livestock, according to Ed Bangs, the Gray Wolf Recovery coordinator.

Meanwhile the only options for ranchers are penning animals at night and hiring extra hands to monitor herds.

Reports are surfacing that deer are being driven out of the forests and into the towns. They regularly eat vegetables, fruit and berries in backyard gardens in Helena.

Young elk are being destroyed in Yellowstone Park, according to reports. Only 100 permits for hunting elk are being issued this year in the Beartooth Game Range compared to 400 to 500 a few years ago.

If the deer and elk become less plentiful, the multiplying predators are going to be looking for more variety in their diets— perhaps red and black angus cattle?

City folk, whose only contact with wild animals is in zoos where they are behind bars, need to become aware of the danger they represent where they are not in cages. Many have grown fond of "Smokey the Bear" and friendly cartoon and storybook bears.

They sometimes, unknowingly, place themselves and their children at risk by providing food for wild bears. Recently a young bear in the Adirondacks, in New York state, snatched a baby from its carriage and took off into the woods. The baby died.

Bears easily lose their fear of humans and become aggressive when they see an opportunity to obtain food. In Montana, the government has placed the grizzly bear at the top of the food chain instead of man.

Some parents in northwestern Montana believe, with good

reason, that it is unsafe to leave their children unsupervised in their own back yards if they live on the edge of town.

Dr. Arthur Robinson reported, in a recent issue of his newsletter, *Access to Energy*, that there is an environmental plan to replace 17,000 people who live in a valley in Oregon with bears and wolves.

Who is voting for this?

TRANSPLANTED WOLVES IN THE WEST

A dangerous development which mushroomed across many European countries is now taking hold in the United States. Most of this country's forests and wildlife that once belonged to citizens is fast becoming the property of the state.

This is dangerous for residents of the northwestern United States where federal conservation efforts have now taken on life-threatening implications as predator species are "reintroduced" to national parks and neighboring areas.

Mountain lions, grizzly bears and wolves are protected by federal laws. The Wyoming legislature, however, recently rebuffed the federal government, declaring wolves to be predators because of the damage that they are doing.

Neighboring Montana disagrees, arguing that federal law supercedes state law.

Hunters, in the form of license fees, and taxpayers have contributed millions of dollars in order to provide for the management of deer and elk, so that they would be available to future generations of hunters.

Unfortunately, the protected, hungry predators are to be the major beneficiaries of these investments.

In Montana, most of the wolves are in the mountainous western part of the state. It is a matter of time before they spread to the eastern parts where most of the cattle ranches are located. Ranchers are well aware of their potential losses when there is an invasion of wolves.

There is increasing resentment in western states where people have spent a lot of time and money eradicating the very predators that the federal government is now working to reintroduce and protect.

WOLVES PREY ON SHEEP

In Mill Creek, Mont., rancher Jim Melin recently found 15 of his sheep dead with their rib cages torn out and wolf tracks in the snow.

A representative of the wolf recovery team told Melin that sheep, as well as his working sheep dogs—which also provided security for his four small children—were attractive targets for wolves. He considered the offer of a permit to kill one wolf, if he saw one on his land in the next 45 days, to be inadequate.

Joe Fontaine, a wildlife biologist, explained that wolves need more food this time of year because their pups are growing and eat more.

If it is necessary to ensure the survival of wolves in the United States, Montanans ask, why limit them to the West? Perhaps there should be a pack in Central Park for the edification of New Yorkers and one in Rock Creek Park for Washingtonians.

It would be a salutary experience for more Easterners to observe how these animals can adapt to the available food supply. Of course a few dogs and cats may turn up missing, but some believe there are too many of them around now turning feral because of neglect.

GUN OWNERSHIP & CRIME

Information on the effects of Britain's disarming of its law-abiding citizens previously published by this newspaper is important enough to bear repeating and warrants further comment.

Stay away from London. You are six times more likely to get

mugged there than in New York City.

This is the conclusion one can draw from author of *Guns and Violence: the English Experience*, Dr. Joyce C. Malcom, who has provided information about the results of gun control which should make us all think.

English police used to carry only a truncheon. In the past, these unarmed police officers sometimes borrowed guns when they needed them to enforce the law.

Now, however, they are starting to carry guns because of the prevalence of gun crimes. And they are seeking advice from the New York City Police Department.

What is behind this recent upswing in violent crime plaguing British towns and cities? A plethora of laws which make it a crime for law-abiding citizens to defend themselves or to injure criminals engaged in their chosen profession.

The 1953 English Prevention of Crime Act made it a criminal offense to carry any article, in a public place, that could be used for self-protection. A string of recent arrests indicate that the British are zealously enforcing this law.

Eric Butler, a British Petroleum executive, was recently arrested for carrying a walking stick enclosing an ornamental sword blade. Butler had used it to fight off two thugs who attacked him.

Also, an elderly lady was arrested for scaring away a group of young delinquents menacing her by firing a cap pistol at them.

The British government appears to be more interested in preventing criminals from being hurt than in preventing crime. Few criminals are jailed. They are usually fined. According to Dr. Malcom, not even violent crimes draw severe penalties.

In addition, there seems to be more concern for the welfare of criminals than the citizens.

The case of Tony Martin illustrates this point. He shot and killed one burglar and wounded another one in the course of the seventh robbery of his farm.

As a result of his actions, Martin was sentenced to life in prison for killing the one, and an additional 10 years for wounding the second and 12 months more for possessing an unregistered shotgun.

The wounded burglar, a man with a long criminal record, was sentenced to three years in jail, but released after 18 months.

The thief is suing Martin for causing an injury, which kept him from working—with legal assistance from British taxpayers.

Cui Bono? Who is benefiting from Britain's gun laws?

The answer seems clear. It is not law-abiding citizens. It is criminals.

These latest examples make it appear as though there is a criminal lobby helping English politicians write their laws.

It is worth considering whether something similar is funding the efforts of gun control advocates in this country. Certainly the drug lords have the money to finance such activity. Or perhaps it's the trial lawyers who specialize in defending criminals.

Perhaps the motivation for gun control is to make the world safer for politicians.

In this regard there is an example of the law of unexpected consequences which should be considered.

At the now famous battle of Agincourt, the French king lost the flower of French chivalry because he hired Italian archers, armed with crossbows, who proved to be no match for the English archers, who were armed with longbows.

English archers slaughtered France's knights.

French kings made it illegal for their common people to own bows and arrows because they feared these weapons might be used on them.

In contrast King Henry VIII had sponsored archery contests and made almost every other sport, including tennis, illegal.

Politicians should remember that Machiavelli concluded, long ago, that there is no way for even a prince to protect himself from a determined assassin, and that his best assurance of safety resides in his people's love and respect.

Switzerland, the only country whose people have maintained their freedom for 700 years, encourages small arms marksmanship by subsidizing ammunition purchases and sponsoring matches. Almost every home has an assault rifle in easy reach and someone ready and able to use it. Perhaps this is why Swiss crime rates are very low. In a country with seven million people, only about 100 homicides occur annually.

Criminals in England have benefited from the disarming of law-abiding citizens. Is it reasonable to believe that Americans will not have a different experience if we allow gun control advocates to disarm law-abiding citizens?

GÖTTERDAMERUNG FOR THE LONE SUPERPOWER

Is America, the lone superpower, facing its Götterdamerung —its twilight of the gods? China, which Napoleon advised everyone to let sleep because when she woke she would shake the world, is no longer sleeping and she is now part of a potent alliance. On Nov. 10, 2004, Russian President Vladimir Putin gave notice to the world that the formerly sleeping giant was now a member of an association that contains three quarters of the people, 80 percent of the natural resources and the majority of the scientists and engineers of the world.

The new coalition included India, China, Russia and Brazil.

Six days after Putin's announcement of Brazil's joining the club, that country was visited by International Atomic Energy inspectors. On Nov. 24, 2004, Brazil had a permit to start the early stages of uranium enrichment program.

Not too long afterwards, an ineffective attack on Venezuelan President Hugo Chavez's barracks, which Chavez blamed on the CIA, resulted in that country joining the coalition.

The Monroe Doctrine, promulgated to keep European influence away from South America, has been breached from another quarter. It may have become an anachronism.

A few days after the Venezuelan fiasco, Russia agreed to supply Venezuela with 50 of its most up-to-date fighter aircraft, MIG 29 SMTs. All are to be painted blue, the proper color for low level attacks over water. Twenty of them are to be equipped to launch Onyx ram-jet anti-ship cruise missiles. These are said to weigh over two tons and have a terminal velocity of 2,460 feet per second.

It is claimed that they can knock out a tanker or an aircraft carrier with a conventional warhead.

The Venezuelans have reported that the fighter jets will be used to protect the Panama Canal. The canal is now operated by a Chinese company which many contend is connected to that country's military.

In order to train Venezuelan fighter pilots, the Cubans have been provided with five of these modern fighters which will supplement their less modern MIGs.

There are reports that former Russian intelligence operations on Cuba have been taken over by the Chinese. The Chinese are also building a container port in the Caribbean which, of course, should be well situated to supply logistical support for Cuba and Venezuela.

China, whose GDP has been growing phenomenally for several years, is expected to grow at about 8 percent per year possibly for the next 15 years. In 2004 it is reported to have used 66 percent of the world's iron ore, 40 percent of its steel, 30 percent of its coal and 55 percent of its cement. There are now some 7 million automobiles in the hands of the Chinese people up from 700,000 in 1993.

Its economic growth is based on low cost labor.

The lowest paid workers in their factories work a 14-hour day reportedly for $100 per month. This has enabled it to monopolize many industries.

It is a major supplier of products globally, and its impact on markets all over the world is substantial. Presently 100 percent of Levi's blue jeans, 60 percent of cell phones, 50 percent of shoes and more than half of the TV sets are made in China.

In the United States, Americans are dependent on China for 95 percent of their shoes, 90 percent of their sporting goods and 80 percent of their toys.

CROP CIRCLES

The primary goals of the BLT Research Team are the discovery and scientific documentation of physical changes induced in plants, soils and other materials at so-called crop circle sites by the energy responsible for creating them and to determine, if possible, from this data the specific nature and source of these energies.

Nancy Talbott, president of BLT Research Team, Inc., reported to attendees at the Ozark UFO Conference on April 11, 2003, at Eureka Springs, Ark., that she had observed a crop circle made in a field very close to the house where she was staying with friends in Holland.

She claimed she was alerted to the possibility that something unusual might be about to happen because cattle nearby, usually very quiet, started bellowing while she was preparing to go to bed about 3 a.m.

Since this kind of animal activity has been reported to precede the appearance of crop circles she glanced through a window, which overlooked the nearby fields. She saw what she characterized as a cylindrical plasma discharge hitting the ground.

This was repeated a second and third time to form a crop circle by flattening string bean plants.

One of her associates also observed this phenomenon from a different location in the house.

No sound, odor or electrical tension was observed during these apparent discharges.

She was unable to find the source of the discharges. The sky was overcast at the time.

Talbott has been investigating crop circles since 1992. BLT claims that several hundred trained field-sampling personnel collect plant and soil samples in the United States, Canada and Europe for analysis.

This not-for-profit corporation has a large database, which, hopefully, can be used to significantly advance scientific understanding of the unusual phenomena involved in the formation of certain crop circles.

William C. Levengood, a biophysicist with 50 scientific papers to his credit, did most of the analysis of plants and soils.

Levengood found changes in the plants and the presence of tiny iron "beads" in the soil samples, which he interpreted as indicating that a considerable amount of energy is involved in the formation of the circles.

Previously microwave energy was assumed, but Talbott says her observation of unusually massive discharges implies a much higher level of sophistication, which, she believes, appears to be beyond present known earthly capabilities.

BLT research papers have been published in *The Journal of Scientific Explorations* and *Physiologia Plantarum*, a Danish publication.

A paper involving a study of crop circle soils by Dr. James Reynolds is being prepared for submission to *Nature*.

Additional information is available on the news web site, rense.com.

HELP FOR THE AGING BRAIN

Age-associated memory impairment, experienced by almost everyone over 50, is considered to be a normal sign of aging. It is one of the problems senior citizens are supposed to accept along with weight gain, hair and muscle loss and low or no sex drive.

Forgetting a telephone number in the middle of dialing or

locking your car door with the keys inside and the motor running are ascribed to "senior moments." Loss of brain capacity due to decreasing numbers of neurons in the aging brain is the accepted cause of this condition. There are said to be about 12 billion neurons in the brain of a healthy young adult, which disappear at the rate of 100,000 every day.

While older people may find senior moments acceptable, particularly if they have a sense of humor, most have an understandable fear of senile dementia and Alzheimer's Disease (AD). Rarely encountered in the past when life expectancies were lower, AD now affects about 50 percent of Americans over 85, AD takes about 20 years to develop and leaves people in a pathetic condition capable of surviving another 10, at tremendous cost to themselves, their families and society.

While conventional medicine has little to offer older people with brain impairments, some unconventional physicians are reporting progress.

Using a blend of Eastern and Western medicine, Dr. Dharma S. Khalsa of Tucson, Ariz., is one of them. His book, *Brain Longevity*, describes an integrated program of diet, lifestyle changes and medicine designed to revitalize aging brains. He claims that it can diminish the effects of age-related memory deficits and slow the progress of AD in its early stages.

He believes that besides loss of neurons and occluded arteries, a major cause of brain deterioration is chronic exposure to toxic levels of the hormone cortisol, a product of the adrenal glands that sit on top of the kidneys.

Animal studies conducted by Robert M. Sapolsky of Stanford University have demonstrated how cortisol damages brain cells. First, it inhibits the utilization of blood sugar by the brain's primary memory center, the hippocampus. Second, it interferes with the functioning of the brain's neurotransmitters. Third, it disrupts brain cell metabolism, leading to destruction by free radicals. Other researchers have confirmed that cortisol has equally bad effects on human brains.

When we are subjected to stress our adrenal glands release adrenaline, which prepares us for "fight or flight." Blood sugar rises, hearts beat faster, arteries constrict and digestion slows as a result. If the stress continues, they secrete cortisol, which maintains a long-lasting stress response. Long-term excessive stress causes the "general adaptation syndrome" described by Dr. Hans Selye in his book *The Stress of Life*. This syndrome with chronic oversecretion of cortisol adversely affects the brain as well as emotional, intellectual and physical health. There is a part of the brain that can shut off the flow of cortisol, but this deteriorates with age and excessive use.

If one can manage or "let go" of stress, its impact is minimized. Social support systems can reduce its effects. On the other hand, effects are magnified if it's out of control.

Dr. Herbert Benson of Harvard described the physiology of stress and its effects on degenerative diseases in his book, *The Relaxation Response*.

When invoked, the response increases blood flow to the brain, decreases blood pressure, reduces oxygen consumption, relieves muscle tension, energizes the immune system, decreases cortisol levels and potentiates memory. A flood of calming neurotransmitters is said to be responsible.

Besides medications, stress reducing activities and physical mental and mind-body exercise, Khalsa prescribes a heart-healthy diet and concentrated nutriments in supplement form.

The antioxidants are featured because free radicals can damage brain cells. Choline from lecithin, a precursor of acetylcholine, a neurotransmitter involved in the establishment of memories, is particularly emphasized. This substance also heightens mental ability in healthy people. With vitamin B-5 and DMAE, it generates acetycholine, necessary for establishing memories in the brain. Phosphatidyl serine, abundant in brain cell membranes, improves cognition and potentiates acetyl carnitine, which improves communication between the two sides of the brain.

Gingko biloba enhances cerebral circulation and increases cognitive ability.

Approximately 600 mg. was found to improve short-term memory an hour after ingestion in one test, although 40 mg. had no effect. It has been found that 360 mg. per day has substantial beneficial effects on nursing home patients. Last but not least he recommends ginseng, a powerful adrenal gland tonic, which has a long history of use as an adaptogen in Oriental medicine.

Many elderly patients are hormonally deficient, and all should be tested for dehydroepiandrosterone (DHEA) deficiencies.

Pregnenelone, sometimes referred to as a neurohormone because large amounts are synthesized from cholesterol in the brain and peripheral nerves, is the raw material for the other steroidal hormones. Depressed patients typically have a low level of pregnenelone. Although most effective in women, this material can improve memory and mood, does not require a prescription and is reported to have no significant side effects.

Melatonin is secreted by the pineal gland and helps control sleep cycles. Production declines with age, and supplementation helps up to 80 percent of people with insomnia. It can help restore normal sleep patterns in AD patients, which helps them function better, and it reduces the number of late night disruptions their caregivers have to cope with.

Human growth hormone (HGH) may be appropriate for early AD, but it is very expensive, and research on it is still incomplete. Because hormones have profound effects on the body, knowledgeable physicians should supervise patients who take them.

Dr. Abram Hoffer, a psychiatrist, the author, with Dr. Morton Walker, of *Smart Nutrients*, has developed an orthomolecular based anti-senility program. His list of primary degenerative brain disorders includes AD, Creutzfeld-Jakob Disease, Huntington's disease, Pick's Disease and senile dementia.

There are no nutritional treatments for Creutzfeld-Jakob Disease and Pick's Disease, and Dr. Hoffer's AD patients did not respond to his nutritional therapy. However he has successfully treated senile dementia patients nutritionally and, in one case, was able to reverse the effects of Huntington's disease with a combination of vitamins B, C and E.

His first experience with curing senile dementia involved his mother. Just before leaving for Europe in 1954 he found that she showed definite signs of dementia for which he "knew" he had no cure. A placebo was the best he had to offer, and he chose niacin, which he had been using with patients for two years.

Much to his surprise, about six weeks later, he received a letter from her reporting that she was feeling much better, her vision was restored, neuralgia was gone, memory was normal and the small arthritic bumps on her knuckles were disappearing. He was amazed at the power of the placebo effect, which, he felt, had clouded his mother's assessment of her health. Besides he knew that Heberdon's nodes, those bumps on her knuckles, never go away.

When he returned from his three-month tour of Europe and visited his mother he was astonished to find that mental and physical responses showed that the deterioration that he had previously observed was now reversed. Even the Heberdon nodes were gone.

Subsequently his mother enjoyed good mental health and led an active, productive life until she had a stroke and died 21 years later at the age of 87. During those 21 years she took one to four grams of niacin every day plus vitamins C and E and other nutrients.

CHELATION THERAPY

In *Smart Nutrients* Hoffer mentions the case of a 70-year-old man exhibiting all the symptoms of AD, including inability to speak, who apparently recovered. The patient stated that he had no recollection of the condition, but that halfway through a course of 20 chelation treatments he woke up. Thus AD may represent the effects of an oxygen-deprived brain, which has fallen asleep.

Dr. Edwin Boyle at the Miami Heart Institute investigated the effects of hyperbaric oxygen on patients who were becoming senile as well as some who had been senile for several years. His patients received five half-hour treatments a week for two weeks in a chamber, which contained pure oxygen at two atmospheres of pressure.

In some cases hopelessly senile people began to act normally. Unfortunately all patients returned to their former condition after a few months. It is interesting to speculate about what might have resulted if Boyle's patients had been given a course of chelation treatments in addition to the hyperbaric oxygen.

Chelation involves infusion of the bloodstream with ethelenediaminetetraacetic acid (EDTA). This is known to react with heavy metals and facilitate their removal from the body via the kidneys.

The FDA approved the use of this material for this purpose. It was used to cure plumbers and painters of lead poisoning. As a side effect it also benefited their cardiovascular systems. It is believed that EDTA sequesters calcium, which anchors plaque deposits in arteries, thus facilitating their removal by normal blood flow.

A retired judge of the Court of Claims said that he had been told that one of his feet had to be removed because of diabetic problems 30 years ago. Instead, he took about 30 chelation treatments and saved his foot. Besides, he said, "After 10 treatments my brain opened up."

Now, he returns for chelation treatment every six months to the physician who saved his foot. A friend whose left carotid artery was over 80 percent occluded found that occlusion was reduced to 50 percent after 10 chelation treatments.

B VITAMIN DEFICIENCIES

Deficiencies of some of the B vitamins can cause psychiatric diseases, for example thiamine and niacin.

Lack of sufficient thiamine in the diet causes the psychiatric and neurological symptoms of a disease called beriberi. Alcohol users are prone to develop Korsakoff's syndrome, a memory disorder caused by thiamine deficiency. Excessive consumption of sugar can also cause thiamine deficiency. In older people this syndrome can easily be misdiagnosed as senile changes.

Niacin deficiency causes pellagra with symptoms of depression, anxiety and fatigue in early stages leading to an organic psychosis resembling senile psychoses, typically schizophrenia.

Hoffer recommends that older people take the following anti-senility vitamins in the amounts indicated:

- Thiamine . . . . 250 mg./day in divided doses
- Niacin . . . . 1000 mg./after each meal
- Pyridoxine . . . . 500 mg./day
- Pantothenic acid . . . . 250 to 750 mg./day
- Vitamin C . . . . 3 to 9 g./day
- Vitamin E . . . . 800 to 1,600 IU/day

His book also offers valuable information and advice about major and trace mineral needs, foods and food allergies. The last chapter recommends exercise and describes advantage to be gained by religiously adhering to a well-chosen program.

RESEARCH ON POWERS OF THE MIND

The federal government and universities throughout the nation are exploring the human mind to determine if the power of positive thinking can move mountains.

Scientists have long been aware that the mind is many-faceted. In double-blind studies, patients receiving a placebo, or phony pill, often do almost as well as those taking the real medication. Their minds convince them they are being treated and they respond.

Mothers exhibit uncanny sensitivity when danger threatens an absent child. Dowsers are able to determine where to drill for water hundreds of feet below in solid rock. Anecdotes abound.

About 1970, a young man said he could turn a light bulb on

and off with brainpower alone. With a couple of wires attached to his head, and a little black box, he demonstrated this to about 100 people at NASA headquarters in Washington.

Now there is an ongoing $24 million Defense Advanced Research Projects Agency program to determine what can be done with the power of thought and modern technology.

The Technology Review May 2003 issue provides an overview of what is happening in this area of research and development.

Ted Berger, a neurobiologist at the University of Southern California, is attempting to map the circuitry involved in memory inside a rat's brain. The objective is to gain an understanding of how minds and machines can interact. If successful, the result could be new techniques for controlling machinery and vehicles with the power of thought and, possibly, the wireless communication of thought between people.

At Duke University in Durham, N.C., Miguel Nicoleli has a rhesus monkey learning to control a robot arm through brain signals picked up by an implanted electrode. At the University of Michigan Daryl Kipke is attempting to teach rats and monkeys how to control six-legged robots. He hopes to have a monkey in St. Louis navigate a robot through a maze in Ann Arbor via the Internet this summer.

OTHER STUDIES

Richard Anderson at Caltech is developing electrode systems for use in brain research. John Donohue at Brown University is developing prostheses to permit paralyzed people to interact with computers. Andrew Schwartz at the University of Pittsburgh is working on neural prostheses. Harvey Wiggins with Plexon in Dallas is developing hardware and software for recording and analyzing brain signals.

The Ozark Research Institute in Fayetteville, Ark., has been studying the activities of brainwaves during remote psychic healing. The book Infinite Mind by Dr. Valerie V. Hunt describes some of her work in this area at the University of California. The National Institutes of Health recently supported an investigation of psychic healing of brain tumors.

The American Society of Dowsers has been a center for individuals engaged in unsponsored research projects on dowsing phenomena. In Germany and Russia government programs have supported investigations involving dowsers. French and Egyptian dowsers recently investigated ancient Egyptian dowsing activities.

Many who deal with psychic phenomena believe the brain to be a computer and that thought originates in the soul. They also believe that energy, not presently recognized by scientists, makes the power of thought possible. Modern physicists have said that, after many years of studying nature, they have arrived at conclusions reached thousands of years ago by theologians. Perhaps brain research will reach a similar juncture.

GOD IS NOT DEAD

At the annual meeting of the Ozark Research Institute (ORI) April 27, Larry Dossey, M.D., told members gathered at the Mount Sequoyah Methodist Retreat Center in Fayetteville, Ark., that gold and possibly platinum standards had been used to obtain proof of the efficacy of long distance psychic healing. The modality used in the studies discussed by Dr. Dossey was prayer.

Many different organized religions and even groups without religious affiliation have been involved in these studies. Positive results were obtained by many different groups and no single religion appears to have a monopoly on the power of prayer. That long distance psychic healing works was not news to the 3,000 members from 13 different countries which make up ORI. The group has been gathering data on "miraculous cures" since December 1992 when it was chartered by the state of Arkansas to study the powers of the mind. The group's president, Harold

McCoy, is a retired military intelligence officer who is now an ordained minister.

One of the first researchers to investigate links between cancer and the mind was Lawrence LeShan who wrote a seminal book on distance healing, *The Medium, The Mystic, and the Physicist*, which was published in 1996.

Bernard Grad of McGill University, Montreal, Canada showed that thoughts and intentions affect living things in medically significant ways such as the healing of wounds and the growth of tumors.

Grad also proved that changes observed were not due to the placebo effect by experiments on seeds, plants and mice which could not be expected to be susceptible to suggestion. *The International Journal of Parapsychology*, in its third issue for 1961, contains an article which Grad wrote with collaborators R.J. Cadoret and G.I. Paul from the University of Manitoba titled: "The Influence of an Unorthodox Method of Treatment of Wound Healing in Mice."

In December 1996 a conference on "Intercessory Prayer and Distant Healing: Clinical and Laboratory Research" was held at Harvard University to report on and discuss experiments designed to test whether or not individuals could mentally heal distant persons who were unaware of the intervention. The Harvard meeting was convened by Marylin Schlitz, the research director of the Institute of Noetic Science, which had been founded by Edgar Mitchell, the Apollo 14 astronaut who had walked on the moon.

Mitchell himself conducted a thought transmission experiment from his spacecraft using a table of random numbers and so-called "Zener" symbols made popular by Dr. J. B. Rhine of Duke University.

Mitchell reported that the results obtained were significantly better than could be expected by chance.

It is an accepted fact that expectations and positive thinking of a patient can alter perception of pain and response to medications beyond the placebo effect. But expectations and attitude of the caregiver may also play an important therapeutic role. Studies of brain wave behavior during psychic healing also indicate that the healee's brain waves are affected by the healer. Dr. Valerie V. Hunt's book, *Infinite Mind*, contains reproductions of electroencephalograms which illustrate this point. Electroencephalograms are weak electrical signals obtained from electrodes placed on a person's head that signal the state of brain cell activity. They were obtained at her Energy Fields Laboratory at the University of California.

Although limited in scope at this time, ORI's research indicates that a healer using visualization techniques has brain wave energy distributed across all levels of normal activity with greatest concentration in low frequencies corresponding to the sleeping state. Thus the healer's brain appears to be functioning in the alpha, beta, gamma and delta modes simultaneously. This writer, personally, knows a woman who was discharged from Georgetown University Hospital with an inoperable torn heart valve in 1995. She is, however, still alive and her present physician finds the organ healed.

Two members of ORI worked on her remotely soon after her discharge from Georgetown. Three dowsers were able to obtain indications of positive changes in the condition of her heart during the two periods of psychic healing.

Evidently there are some things which still defy rational scientific explanation. Perhaps, after all, the reports of God's death are gross exaggerations.

POSTSCRIPT

THE ICEBERGS OF EASTERN GREENLAND

For over 20 years "global warmers" have been predicting dire man made catastrophes, including the flooding of New York City caused by melting of the world's stock of ice. Satellite photography leaves no doubt that sea ice around Greenland and the Antarctic peninsular has melted. Also reports indicate that many temperate zone glaciers and some arctic zone glaciers are melting. The plight of polar bears and the tundra in arctic regions also supports their case. On the other hand, satellite radar surveys show that the ice caps of Greenland and Antarctica, which together contain 95 percent of the world's ice, haven't stopped growing. This supports the contentions of those who believe that the current interglacial period is coming to an end from natural causes as have innumerable others during the last 20 million years.

Who do you believe? The big money is supporting the global warmers even if the predicted catastrophes keep being pushed off into the future. But thousands of scientifically trained Americans are not convinced.

To satisfy my curiosity I went on an expedition to Greenland to see what life was like in the arctic and to have a look at the ice.

Greenland, a Danish possession, is the largest island in the world with an area of 2.2 million square kilometers. About 5% of the world's ice sits in a cap, which covers it's entire interior. Only 410,000 square kilometers on the coast is not under the cap. In some places the cap is over 3000 meters, almost 10,000 feet, thick. Greenland's northern extremity is the closest land to the North Pole. Water from the Arctic Ocean flows southward along the east coast carrying icebergs from Greenland to help cool the North Atlantic Ocean.

The expedition visited Scorsbysund, the world's largest complex of fjords, about halfway up the eastern coast, I had an opportunity to observe glaciers and icebergs first hand. A few hours in Ittoqqortoormit, about 71 degrees north, provided a picture of life above the Arctic Circle. This municipality, first settled in 1925 contains about 500 people. It is about the size of Great Britain and is relatively close to the largest national park in the world, encompassing about a third of the icecap. The park is not open to the public. Only permanent residents of Ittoqqortoomit and Qaanaaq, another town, are allowed to enter without a permit. Allegedly there are many birds and mammals in the park. Our group saw a few musk oxen, an arctic fox and a few seals in the fjords.

The people I saw during my visit, both children and adults, appeared to be happy and active. One man I met on the gravel road through the town greeted me with a wide grin and a handshake. He spoke perfect English although Greenlandic, an Inuit language, is taught in the schools along with Danish. He and the other people I saw appeared to be Inuits, relatively small dark haired people with the reputation of being extraordinarily adaptable. There were 50 or 60 small houses in the town some with a little greenery around them. One house had a polar bear skin hanging over its porch railing. Hunting and fishing are primary activities. One beach, which we visited, had the remains of about 10 narwhals, which these villagers had killed. We were told that the hunters were going back for the rest of the meat having been unable to transport all of it at once.

I also saw some dogs chained in a line along side of a small stream, which passed through the village. They howled for a while at the sight of new people, but they quieted down quickly. Two or 3 pups ran free in the town, but no mature dogs. These are working dogs, not pets, and are required for transportation during the winter months. The few roads I saw were unpaved and a couple of all terrain vehicles passed me as I walked up to the local weather station. There were no sidewalks.

There are no Trump Plazas in Ittoqqortoormit, no football stadiums or baseball fields. But there are plenty of majestic snow covered mountains and icebergs in the vicinity. Although remote, it can be reached by helicopter from airfields on the west coast. Dog sled tours are available and other activities are mentioned on the website www.nonnitravelgreenland.com.

A supply ship from Denmark shows up once or twice a year with necessities not produced locally. These are sold in a local store. To keep thing going Denmark provides Greenlanders with a subsidy of about $8,000 per person. To reduce costs some of the inhabitants in other parts of the island have been moved into high-rise apartment buildings. This has caused problems because sled dogs are not well adapted to apartment living.   .

I noted that some glaciers near the mouth of Scorsbysund and the open ocean had melted away, leaving beds of stones at the water's edge. And I saw some that were in process of receding into the cold and forbidding hills and mountains close to the water. However, as the ship I was on, *The Professor Molchanov*, a former Russian Artic research vessel, sailed deeper into the Sund, away from the sea and towards the central ice cap, I saw many glaciers that were not melting, and lots of icebergs which growing glaciers produce. One glacier I saw was 6 miles wide. I understand that there is a 60-mile wide glacier on Greenland, but I did not see that monster.

The icebergs I saw came in many sizes and shapes. There were, of course, many small chunks of ice in the water. However there were a lot of large bergs. Inspecting a group of these in a rubber Zodiak at close range is awe-inspiring. I saw several that were 30 to 50 stories tall. Someone in the party estimated a height of 80 meters for one of them. Close to one of these mammoths I could see its foundation deep under the water. Most of the ice, about 66% of the berg, is submerged as it floats on the surface.

These icebergs are part of the Earths conveyor system, which is responsible for the movement of air and ocean currents that influence weather conditions. As the icebergs from the Polar Regions melt, the cold water they generate travels towards the equator. In the equatorial regions heat from the sun is more intense than elsewhere and it produces both currents of warm water and clouds of water vapor, which travel towards the poles. Some of the water vapor is deposited as snow on the ice caps and subsequently turns into ice. The heat released by the conversion of water vapor into liquid water, snow and ice in the Polar Regions is mostly radiated into outer space. This process creates very low temperatures in the icecaps, for example $-70*C$ has been found in Greenland.

"The Physics of Glaciers" by Patterson states that the maximum temperatures of the Holocene, the epoch in which we live, occurred about

5,000 years ago and that the earth has been cooling since then. BGR, a meteorological research institute in Germany, prepared a chart showing the variations of Greenland temperatures for the last 1,000 years. The highs and lows of these temperatures have continually decreased during this period. These facts indicate that the icecaps in Greenland and Antarctica have been growing for a very long time as the earth has been cooling since the Holocene maximum.

Archeological evidence from Greenland also indicates that ice must have been forming for a very long time. In contrast to experience in the Middle East and elsewhere, evidences of the oldest civilizations are located above those of the newer ones. You have to dig down in Egypt and Iraq to uncover the past, but you have to look at higher levels in Greenland because the central icecap limited sources of food and transportation to coastal areas. Carbon dating of artifacts from the oldest civilizations on Greenland gave their age as 2600 BC. This is 4600 years ago, near the time of the Holocene Maximum. These remains were found 12 meters above present sea level.

Assuming that the open areas of oceans and fresh water lakes remained constant for 4600 years, it is possible to obtain an estimate of the water, which has been removed and deposited as ice. This amounts to over 9 million millon tons or, roughly 2 billion tons of ice per year. If the increasing weight of the ice has caused the land to sink, and there are good reasons, in the form of discrepancies between ancient and modern maps, to believe that it has, then my estimate is very conservative.

Satellite radar surveys of Antarctica and Greenland show that the icecaps in both places are growing. Reports from the Mohn Sverdrup Center for Global Ocean studies in Norway state that the Greenland ice cap has increased over 21 inches in the last 11 years. A recent paper published in Science claims that ice is being deposited at the rate of 26.8 billion tons a year in Antarctica.

Numerous newspaper, Internet and TV reports claim that global warming is causing extensive melting of glaciers and permafrost, which might lead to drastic increases in sea level threatening inundation of New York City for example. On the other hand there is no evidence that the sea level is actually increasing, only forecasts by computer programs. . The melting observed in Antarctica is largely confined to the Peninsula area, which juts out into the ocean. Winds from the ocean, which is much warmer than the icecap, are providing the necessary heat. Other parts of Antarctica are getting colder and there the ice is depositing. . Photographs of Greenland show clearly that the sea ice surrounding it has been melting, but there is no evidence that the ice cap, where most of the ice is located, is doing anything but growing. Again, as in Antarctica, temperatures on the ice cap are decreasing. In 2003 Edward Hanna and John Capellen reported that coastal stations in south Greenland have recorded decreasing temperatures for 44 years.

As previously mentioned there has been a long-term secular downtrend in earth temperatures since the Holocene Maximum. Superimposed on that long-term trend are shorter-term fluctuations in temperature. In 1000 AD it was warmer than it is now and Northern Europe had a "Golden Age" when the Vikings farmed in Greenland. Their settlements disappeared when the earth cooled from about 1300 AD to 1700 AD, the depths of the Little Ice Age, an historical fact. Man's actions did not cause this. It was Mother Nature's work. Subsequently the earth has warmed for 300 years. It is reasonable to expect some ice to melt after that many years of warming. However, the fact that temperatures now are lower than they were in 1000AD indicates that the long-term trend is still in force. There is no evidence that a mere 300 years of warming has reversed the 5000-year secular cooling trend.

While it is a fact that carbon dioxide increased during the last 100 years as a result of the industrial revolution and increasing population, what caused temperatures to rise during the first 200 years? Mother Nature was at work again! The available data do not support the contention that the minuscule increase in carbon dioxide concentration – from 0.03 to 0.04% of the atmosphere – significantly affected earth temperatures. On the other hand it is easy to find a close connection between cyclical changes in the radiation supplied by the Sun and conditions on the earth. In fact Milutin Milankovic, a Serbian scientist provided mathematical support for the theory that variations in the orbits of Earth. Sun and Moon were responsible for recurrent ice ages without the aid of a computer while incarcerated in a jail cell.

Nature is highly cyclical: night follows day, winter follows summer, global cooling follows global warming, and glaciations follows interglacial periods. All of these cycles can be explained in terms of movements of the earth-moon system around the sun. Dr. Willard Libbey carbon dated material connected with the end of the last glaciation and found that it was 11,000 years old. Other scientists have found that, during the past 20 million years, these periods have never lasted longer than 12,500 years. Therefore it is reasonable to believe that we are coming to the end of our present interglacial period. Sometime, within the next 1700 years, the earth is going to get much colder and less hospitable. The end phases of past interglacials have lasted 75 to 150 years with very rapid changes in the last 25 years.

As the earth receives less warmth from the sun, the increasing mass of the growing icecaps stresses the earth's crust. According to one theory, this increases volcanic activity. Most volcanoes, 80% of the total, are underwater. When they irrupt ocean temperatures increase. Accordingly: (1) air temperatures increase and some temperate zone glaciers melt; (2) sea ice and ice on land near the sea melts; (3) evaporation increases in the tropics and more water vapor travels to the poles where it deposits as snow. When, as and if this volcanic action diminishes, as the crust adjusts, the earth should cool and the probability of glaciations should increase.

In 1975 mainstream climatologists, the National Academy of Sciences, the CIA and the New York Times reported that this interglacial period would end in the year 2000. The fact that it didn't end proves that weather forecasting is a difficult art. But in 1977 a political decision resulted in a change of focus. A new Academy of Sciences Report, prepared by Dr. Rodger Revelle, one of Al Gore's professors, predicted rapid global warming because of the greenhouse effect of increasing concentrations of carbon dioxide in the atmosphere. Climate research funds became available only for studies of global warming. However ubiquitous water vapor is a far more important greenhouse gas than carbon dioxide. Professor Richard Lindzen, a climatologist at MIT, claims that 98% of the greenhouse effect is caused by water vapor. There is only a few tenths of a percent of carbon dioxide in the atmosphere and even less of other greenhouse gases like methane

Global warming advocates like Michael Mann generally ignore historical information about the Golden Age of Northern Europe and the Little Ice Age. Their catastrophic forecasts are products of very complex and highly suspect computer programs, having 5000 or more assumptions. They seem to be asking us to forget about the past and to accept the prophecies of new gods, the computers, which only they, the new priesthood, can interpret.

These complex programs provide the basis for the Kyoto Protocol, which many European countries have adopted. The Russians have also adopted it because of the monetary and political benefits to be gained, even though they have expressed doubts about the validity of the underlying science. About 17,000 scientifically trained Americans, including the undersigned, have also gone on record recommending that the Kyoto Protocol be rejected.

In 2003 Senator Inhofe, Chairman of the Senate Committee on Environment and Public Works, stated that the threat of catastrophic global warming was "the greatest hoax ever perpetrated on the American people." In a January 2005 statement he noted that European Environmental Commissioner, Margo Wallstrom, considered the Kyoto objective to be leveling the playing field for businesses worldwide. It is too bad that there aren't more people like Inhofe in Congress!

It seems clear to me that U.S. adoption of the Protocol would result in further de-industrialization and loss of middle class jobs, enhancement of the power and influence of the United Nations and, ultimately, a loss of freedom for Americans.

Perhaps people should be grateful for the present warmth, disregard the politically modified science promoted by the global warmers and pray that the warmth continues. Conditions in the arctic leave a great deal to be desired in my experience.

—JACK PHILLIPS

INDEX

PHOTO SECTION

AUTHOR'S COLLECTION

AUTHOR'S COLLECTION

These pictures were taken in Scorsby Sund on the East Coast of Greenland in September 2005. The glacier shown in the lower picture, which was close to the open sea, was receding. However that shown in the upper picture, closer to the ice cap, was still functioning normally.

AUTHOR'S COLLECTION

AUTHOR'S COLLECTION

The upper picture, also taken in Scorsby Sund in September 2005, shows another functioning glacier plus icebergs spawned by glaciers still closer to Greenland's ice cap. The lower picture shows a Viking ship which sailed from Iceland to New York City during the Bicentennial Celebration. Similar vessels carried Vikings, their families, livestock and possessions to Greenland during the Golden Age of Northern Europe in the 10th Century. They left Greenland when decreasing temperatures, incident to the approaching Little Ice Age, made farming impractical around 1250 AD.

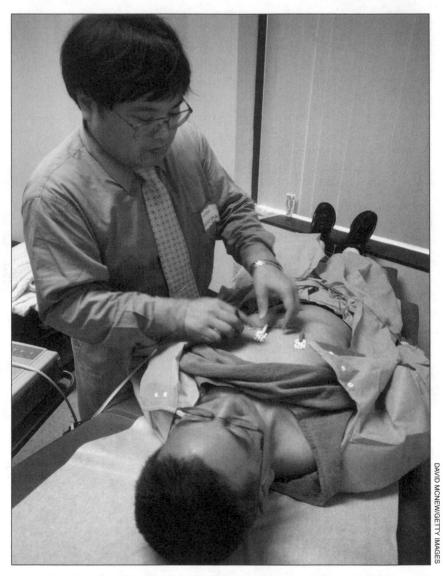

In Los Angeles, Dr. Hideki Mukoyama takes an electrocardiogram (ECG) reading from Toshi Fujishiga, who survived the explosion of an atomic bomb over Hiroshima, Japan by the United States during World War II. A team of physicians from Japan joined Los Angeles physician volunteers examining about 175 survivors of the Hiroshima and Nagasaki atom bomb attacks. This was the 14th visit by Japanese doctors in the 28-year-old Radiation Effects ResearchProject. There are believed to be more than 300 survivors of the attacks now living in Los Angeles.

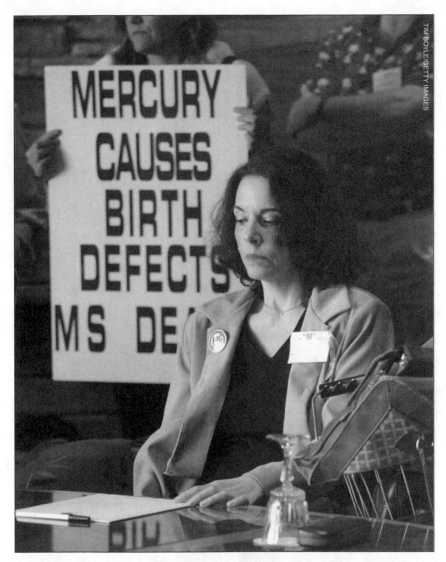

Linda Brocato from Prospect Heights, Illinois, seated in a wheelchair, listens during a press conference September 7, 2001, in Oak Brook, Illinois. Brocato, a coordinator for Dental Amalgam Mercury Syndrome (DAMS) spoke about the acute damage to her health from mercury "silver" amalgam dental fillings. Scientists, dental society representatives, consumer groups and other injured patients attended the press conference where activists denounced the American Dental Association for deceiving consumers by using mercury fillings instead of silver.

Above, a collection of innovative cancer researchers whose efforts were consistently thwarted by the medical establishment. Clockwise from upper left: (1.) Dr. Gaston Naessens (2.) Dr. Royal Raymond Rife in his laboratory. (3.) Nurse Rene Caisse, inventor of the Essiac cancer formula. She was allowed to treat patients, but only if she took no money. (4.) Dr. Wilhelm Reich, who died in prison in 1976. His work is still the subject of great debate.

JEFF TOPPING/GETTY IMAGES

Ronald Cheshier, virology section manager for the Arizona Department of Health Services, holds a tray of blood serum samples that will be tested for the West Nile virus at the department's lab in Phoenix, Arizona. Since it was first diagnosed, thousands of people across the United States have been diagnosed with West Nile Virus and there have been hundreds of deaths as a result of its symptoms such as swelling of the brain. West Nile is a viral disease and high doses of anti-viral vitamin C may benefit its victims.

ATKINS CENTER/GETTY IMAGES

There is no denying that obesity is an epidemic in the United States today. Dr. Robert Atkins, shown above, posthumously received validation for his theories about protein consumption and the relationship between carbohydrates and obesity. See page 124 for a thorough discussion of Dr. Atkins's weight loss program.

The reintroduction of Grizzly bears and Gray Wolves into populated regions of the United States may seem right to environmentalists in the suburbs. But to the folks living in these parts of the country, it makes about as much sense as reintroducing mountain lions into Central Park in New York City or into Rock Creek Park in Washington.

Sure, some crop circles may be the result of pranks created in the middle of the night by college students with entirely too much time on their hands. But, believe it or not, some may be legitimate. For over a decade, a group of prominent scientists have been examining crop circle sites for telltale changes in the in the plants and the presence of metals in soil samples. Above, are these crop circles an elaborate hoax or could they be the result of energy releases from an unknown source.

American Free Press
America's *Last Real* Newspaper

WEEKLY ISSUES of uncensored news from a populist and nationalist perspective. The only newspaper in America written exclusively for the American middle class covering the issues that directly affect America's hardest working, most over-taxed segment of society.

One Year (52 issues plus special health reports) $59
Two Years (104 issues plus special health reports) $99
Trial subscription: 16 weeks for $17.76

Call 1-888-699-NEWS to charge a subscription to Visa or MasterCard. Tell them you saw the ad in *Suppressed Science* and we'll send you a FREE copy of *Dirty Secrets: Crime, Conspiracy and Cover-Up in the 20th Century* by Michael Collins Piper—a $22 value FREE.

When responding by mail, send payment with coupon on reverse to AFP, 645 Pennsylvania Avenue SE, Suite 100, Washington, D.C. 20003. Issues should begin arriving within three weeks of AFP's receipt of payment.